T0313352

Soft and Stiffness-controllable Robotics Solutions for Minimally Invasive Surgery: The STIFF-FLOP Approach

RIVER PUBLISHERS SERIES IN AUTOMATION, CONTROL AND ROBOTICS

Series Editors

SRIKANTA PATNAIK
SOA University
Bhubaneswar, India

ISHWAR K. SETHI
Oakland University
USA

QUAN MIN ZHU
University of the West of England
UK

Advisor

TAREK SOBH
University of Bridgeport
USA

Indexing: All books published in this series are submitted to the Web of Science Book Citation Index (BkCI), to CrossRef and to Google Scholar.

The "River Publishers Series in Automation, Control and Robotics" is a series of comprehensive academic and professional books which focus on the theory and applications of automation, control and robotics. The series focuses on topics ranging from the theory and use of control systems, automation engineering, robotics and intelligent machines.

Books published in the series include research monographs, edited volumes, handbooks and textbooks. The books provide professionals, researchers, educators, and advanced students in the field with an invaluable insight into the latest research and developments.

Topics covered in the series include, but are by no means restricted to the following:

- Robots and Intelligent Machines
- Robotics
- Control Systems
- Control Theory
- Automation Engineering

For a list of other books in this series, visit www.riverpublishers.com

Soft and Stiffness-controllable Robotics Solutions for Minimally Invasive Surgery: The STIFF-FLOP Approach

Editors

Jelizaveta Konstantinova
Queen Mary University of London
United Kingdom

Helge Wurdemann
University College London
United Kingdom

Ali Shafti
Imperial College London
United Kingdom

Ali Shiva
King's College London
United Kingdom

Kaspar Althoefer
Queen Mary University of London
United Kingdom

River Publishers

Routledge
Taylor & Francis Group
LONDON AND NEW YORK

Published 2018 by River Publishers
River Publishers
Alsbjergvej 10, 9260 Gistrup, Denmark
www.riverpublishers.com

Distributed exclusively by Routledge
4 Park Square, Milton Park, Abingdon, Oxon OX14 4RN
605 Third Avenue, New York, NY 10017, USA

*Soft and Stiffness-controllable Robotics Solutions for Minimally Invasive Surgery: The STIFF-FLOP Approach/*by Jelizaveta Konstantinova, Helge Wurdemann, Ali Shafti, Ali Shiva, Kaspar Althoefer.

Routledge is an imprint of the Taylor & Francis Group, an informa business

ISBN 978-87-93519-72-5 (print)

Contents

**4 Antagonistic Actuation Principle for a Silicone-based
Soft Manipulator**

*Ali Shiva, Agostino Stilli, Yohan Noh, Angela Faragasso,
Iris De Falco, Giada Gerboni, Matteo Cianchetti,
Arianna Menciassi, Kaspar Althoefer,
and Helge A. Wurdemann*

PART IV: Human Interface

Krzysztof Lis, Łukasz Mucha, Krzysztof Lehrich
and Zbigniew Nawrat

PART V: Benchmarking Platform for STIFF-FLOP Validation

17 Benchmarking for Surgery Simulators   309

Zbigniew Malota, Zbigniew Nawrat and Wojciech Sadowski

18 Miniaturized Version of the STIFF-FLOP Manipulator for Cadaver Tests   325

Giada Gerboni, Margherita Brancadoro, Alessandro Diodato,
Matteo Cianchetti and Arianna Menciassi

19 Total Mesorectal Excision Using the STIFF-FLOP Soft and Flexible Robotic Arm in Cadaver Models   339

Marco Ettore Allaix, Marco Augusto Bonino, Simone Arolfo,
Mario Morino, Yoav Mintz and Alberto Arezzo

Preface

With the ongoing and insatiably appetite for more advanced surgical tools and a drive to push the frontiers of robot-assisted minimally invasive surgery, the EU funded STIFF-FLOP project set out to challenge the current approach employed to robotically operate on patients. Radically different from the then state of the art, STIFF-FLOP proposed to create and make use of soft and stiffness-controllable robot manipulation tools capable of penetrating into the human body through narrow openings and conduct advanced surgical procedures reliably and in places that could not be reached previously. Due to the predominantly soft nature of the created robot tools, the chosen approach proved to be inherently safe improving vastly over the surgical systems, including the da Vinci Surgical System by Intuitive Surgical which is still dominating the market.

Whilst safety was a paramount element of STIFF-FLOP, there were numerous other advantages of the chosen route to be noted. With its clear goal from the start to explore what nature could offer, bio-inspiration was one of the guiding principles of the project, and we could clearly show that, by using biological role models such as the octopus as a blueprint during our STIFF-FLOP developments, robot tools could be created that are endowed with superior manipulation capabilities in the presence of environmental uncertainty and dynamics and, at the same time, reduced computational burden – a successful instantiation of computational morphology, practically applied!

With a well-defined focus on creating robot-assisted surgical technology, we show that by overcoming the limitations of modern laparoscopic and robot-assisted surgical systems due to restricted access through Trocar ports and the commonly employed straight-line instruments, the STIFF-FLOP robot with its increased dexterity allows operations behind obstacles such as organs or tissue formations and reaching into narrow cavities, opening up avenues to modern patient-oriented surgery. As a direct outcome from using a bio-inspired approach, we were able to develop control algorithms for the surgeon user of our system mimicking the neural structure of the octopus to generate the required control commands for smooth motion primitives executed in the patient's abdomen. Exploiting our biological findings further, we created haptic feedback technology to stimulate the nerves on the user's hands and arms representing interactions between the STIFF-FLOP robot and the soft-tissue environment.

This book will provide an in-depth overview of the research and development activities that were part of STIFF-FLOP and outline the project's achievements and legacy. The work was the outcome from four years of a close-knit collaboration between a team of experts across biology, engineering and medicine. Without the continuous and enduring efforts by all the twelve project partners, the presented achievements would not have been possible. Undoubtedly, STIFF-FLOP has laid the foundation for soft and stiffness-controllable robots as a promising alternative for the current type of rigid-component, straight-line minimally invasive surgery (MIS) instruments. The STIFF-FLOP concept with its innovations in soft material robotics and stiffness controllability is here and it is here to stay. And current researchers including colleagues at Harvard University are taking note, as evidenced by the recent developments of soft robots for MIS coming from their lab. Beyond the novel, bio-inspired thinking in MIS, STIFF-FLOP has many other outcomes to be proud of: Novel, inflatable and stiffness-controllable structures with applications beyond MIS, including search and rescue and human-robot interaction in the manufacturing sector; Miniaturized high degree-of-freedom force and stiffness sensors with high commercial potential in real-world application areas; Novel control paradigms taking bio-inspiration into practical robotic systems; Innovative haptic devices that go beyond the standard joystick approach providing an all immersive body experience; and Improving the understanding of the biological functions in animals such as the octopus provided through the STIFF-FLOP's constructivist approach.

This book consists of five parts. In the first part we present the development of silicone-based stiffness controllable actuators for STIFF-FLOP surgical platform. Specifically, we focus on the technology selection for soft surgical robot, design of manipulator, actuation principles, as well as material selection. The second part of the book is on the creation and integration of multiple sensing modalities, such as force and tactile sensors, as well as sensors for shape detection and bending for flexible manipulator. Part III discusses control, kinematics and navigation for continuum manipulators. Further on, in Part IV, human interface for STIFF-FLOP surgical platform is presented. This includes surgical console, methods of haptic feedback during tissue palpation, as well as haptic feedback sleeves for interaction perception. Finally, Part V presents benchmarking platform for STIFF-FLOP validation, and it is comprised of the chapters on organ benchmarking prototypes, final miniaturised version of STIFF-FLOP manipulator, and the results of surgical validation on benchmarking and cadaver models.

Technical topics discussed in the book include:

- Soft actuators
- Continuum soft manipulators
- Control, kinematics and navigation of continuum manipulators
- Optical sensors for force, torque, and curvature
- Haptic feedback and human interface for surgical systems
- Validation of soft stiffness controllable robots

Acknowledgements

This book is based upon work from STIFF-FLOP project that has received funding from the European Commission Seventh Framework Programme (FP7 – ICT) under grant agreement No. 287728 from 2012-01-01 to 2015-12-31.

We would like to acknowledge a valuable contribution of project partners: King's College London; Universitaet Siegen; Fundacion Tecnalia Research & Innovation; The Hebrew University Of Jerusalem; Fondazione Istituto Italiano Di Tecnologia; Universita Degli Studi Di Torino; Scuola Superiore Di Studi Universitari E Di Perfezionamento Sant'anna; Stichting E.A.E.S; Fundacja Rozwoju Kardiochirurgii Im Prof Zbigniewa Religi; Przemyslowy Instytut Automatyki I Pomiarow Piap; University Of Surrey and The Shadow Robot Company Limited.

List of Contributors

Agostino Stilli, *Department of Informatics, King's College London, London, United Kingdom*

Ahmad Ataka, *King's College London, London, United Kingdom*

Alberto Arezzo, *Department of Surgical Sciences, University of Torino, Torino, Italy*

Alessandro Diodato, *The BioRobotics Institute, Scuola Superiore Sant'Anna, Pontedera (PI), Italy*

Alfonso Dominguez Garcia, *Tecnalia Research & Innovation, Donostia – San Sebastián, Spain*

Ali Shiva, *Department of Informatics, King's College London, London, United Kingdom*

Angela Faragasso, *Department of Informatics, King's College London, London, United Kingdom*

Anthony Remazeilles, *Tecnalia Research & Innovation, Donostia – San Sebastián, Spain*

Anuradha Ranasinghe, *Department of Mathematics and Computer Science, Liverpool Hope University, United Kingdom*

Arianna Menciassi, *The BioRobotics Institute, Scuola Superiore Sant'Anna, Pontedera (PI), Italy*

Asier Fernandez Iribar, *Tecnalia Research & Innovation, Donostia – San Sebastián, Spain*

Ashraf Weheliye, *Department of Informatics, King's College London, United Kingdom*

Daniel Guevara Mosquera, *Center for Robotic Research (CoRe), King's College London, London, United Kingdom*

Dariusz Krawczyk, *Prof. Z. Religa Foundation of Cardiac Surgery Development, Zabrze, Poland*

Emanuele Lindo Secco, *Department of Mathematics and Computer Science Liverpool, United Kingdom*

Erwin Gerz, *Lehrstuhl für Regelungs- und Steuerungstechnik (RST), Department Elektrotechnik und Informatik, Fakultät IV – Naturwissenschaftlich-Technische Fakultät, Universität Siegen, Siegen, Germany*

Giada Gerboni, *The BioRobotics Institute, Scuola Superiore Sant'Anna, Pontedera (PI), Italy*

Helge A.Wurdemann, *Department of Informatics, King's College London, London, United Kingdom*

Hubert Roth, *Lehrstuhl für Regelungs- und Steuerungstechnik (RST), Department Elektrotechnik und Informatik, Fakultät IV – Naturwissenschaftlich-Technische Fakultät, Universität Siegen, Siegen, Germany*

Iris de Falco, *The BioRobotics Institute, Scuola Superiore Sant'Anna, Pontedera (PI), Italy*

Jan Czarnowski, *Industrial Research Institute for Automation and Measurements PIAP, Warsaw, Poland*

Jan Fras, *Industrial Research Institute for Automation and Measurements PIAP, Warsaw, Poland*

Jakub Głowka, *Industrial Research Institute for Automation and Measurements PIAP, Warsaw, Poland*

Jelizaveta Konstantinova, *School of Mechanical Engineering and Materials Science, Queen Mary University of London ARQ (Advanced Robotics at Queen Mary), London, United Kingdom*

Kaspar Althoefer, *School of Mechanical Engineering and Materials Science, Queen Mary University of London ARQ (Advanced Robotics at Queen Mary), London, United Kingdom*

Krzysztof Lehrich, *Prof. Z. Religa Foundation of Cardiac Surgery Development, Zabrze, Poland*

Krzysztof Lis, *Prof. Z. Religa Foundation of Cardiac Surgery Development, Zabrze, Poland*

Łukasz Mucha, *Prof. Z. Religa Foundation of Cardiac Surgery Development, Zabrze, Poland*

Mateusz Macias, *Industrial Research Institute for Automation and Measurements PIAP, Warsaw, Poland*

Marco Augusto Bonino, *Department of Surgical Sciences, University of Torino, Torino, Italy*

Marco Ettore Allaix, *Department of Surgical Sciences, University of Torino, Torino, Italy*

Margherita Brancadoro, *The BioRobotics Institute, Scuola Superiore Sant'Anna, Pontedera (PI), Italy*

Mario Morino, *Department of Surgical Sciences, University of Torino, Torino, Italy*

Matteo Cianchetti, *The BioRobotics Institute, Scuola Superiore Sant'Anna, Pontedera (PI), Italy*

Matthias Mende, *Lehrstuhl für Regelungs- und Steuerungstechnik (RST), Department Elektrotechnik und Informatik, Fakultät IV – Naturwissenschaftlich-Technische Fakultät, Universität Siegen, Siegen, Germany*

Min Li, *Institute of Intelligent Measurement and Instrument, School of Mechanical Engineering, Xi'an Jiaotong University, Xi'an, PR China*

Prokar Dasgupta, *MRC Centre for Transplantation, DTIMB & NIHR BRC, King's College London, United Kingdom*

Simone Arolfo, *Department of Surgical Sciences, University of Torino, Torino, Italy*

Sina Sareh, *Design Robotics, School of Design, Royal College of Art, London, United Kingdom*

S.M. Hadi Sadati, *1. Center for Robotic Research (CoRe), King's College London*
2. Dyson School of Design Engineering, Imperial College London, London, United Kingdom

Thrishantha Nanayakkara, *Dyson School of Design Engineering, Imperial College London, London, United Kingdom*

Tommaso Ranzani, *Department of Mechanical Engineering, Boston University, Boston, MA, United States*

Tommaso Ranzani, *The BioRobotics Institute, Scuola Superiore Sant'Anna, Pontedera (PI), Italy*

Wojciech Sadowski, *Prof. Z. Religa Foundation of Cardiac Surgery Development, Zabrze, Poland*

Yoav Mintz, *Department of General Surgery, Hadassah Hebrew University Medical Center, Jerusalem, Israel*

Yohan Noh, *Centre for Robotics Research, Department of Informatics, King's College London, London, United Kingdom*

Zbigniew Malota, *Prof. Z. Religa Foundation of Cardiac Surgery Development, Zabrze, Poland*

Zbigniew Nawrat, *Prof. Z. Religa Foundation of Cardiac Surgery Development, Zabrze, Poland*

List of Figures

List of Tables

List of Abbreviations

AA	Anhydrous Acrylic Acid
AAm	Acrylamide
APS	Ammonium Persulfate
Cu^{2+}	Copper Anode
DOF	Degrees of Freedom
EDTA	Ethylenediaminetetraacetic Acid
EKF	Extended Kalman Filter
EtOH	Ethanol
F/T	Force/Torque
FCSD	Foundation of Cardiac Surgery Development
FDM	Fused Deposition Modelling
FITC	Fluorescein Isothiocyanate
HD	High Definition
HSV Color Space	Hue, Saturation, Value Color Space
IMSaT	Institute for Medical Science and Technology
LIM	Light Intensity Modulation
LMP	Low Melting Point
M	Mole
MIS	Minimally Invasive Surgery
MRI	Magnetic Resonance Imaging
PDMS	Poly-di-methyl-siloxane
PIAP	(in polish – Przemysłowy Instytut Automatyki i Pomiarów - Industrial Research Institute for Automation and Measurements)
pNaAc	Electro-active Poly (Sodium Acrylate)
pNIPAAM	Poly(N-isopropylacrylamide)
PSP	Point Subjected to Potentials
PWM	Pulse-width Modulation
RCTs	Randomized Controlled Trials
RiH	RobinHeart

RMIS	Robot-assisted Minimally Invasive Surgery
ROS	Robot Operating System
RVIZ	ROS Visualization Tool
STIFF-FLOP	STIFFness Controllable Flexible and Learn-able Manipulator for Surgical Operations
SVM	Support Vector Machines
TEMED	Tetramethylethylenediamine
TIS	Tactile Imaging System
TME	Total Mesorectal Excision
TSS	Tactile Sensing System
UI	User Interface
VCM	Voice Coil Motor
VR	Virtual Reality

PART I

Development of Silicone-based Stiffness Controllable Actuators

1

Technology Selection

**Matteo Cianchetti, Tommaso Ranzani, Giada Gerboni
and Arianna Menciassi**

The BioRobotics Institute, Scuola Superiore Sant'Anna,
Pontedera (PI), Italy

Abstract

The first step for the development of the soft manipulator starts with the definition of both medical and technical requirements. Within these boundaries, the most suitable technological choices have to be taken. Thus, after the manipulator specifications, a survey of candidate actuation technologies is reported. A direct comparison is also provided to highlight the advantages and disadvantages for the specific application field.

1.1 Manipulator Specifications

The flexible manipulator has been designed to meet specific requirements extracted from medical literature and experience and through *in vivo* biomechanical tests of internal organs. On the basis of the limitations underlined by the current robotic instruments used in surgery and the desired characteristics from a clinician perspective, the technical specifications of the manipulator have been derived.

1.1.1 Medical Requirements

In a clinical setting perspective, the STIFF-FLOP manipulator should have the following characteristics, allowing overcoming the main overall limitations of the current available robotic system:

- *Dimensions and maneuverability:* The STIFF-FLOP manipulator should be limited in its overall dimensions and weight, being these directly related to the maneuverability in the operating room. In this context, it is important to bear in mind how standard operating rooms are limited in spaces and generally contain a great number of medical devices, such as the operating bed, the nurse workstation, the anesthesiologist workstation, trolleys, and wardrobes, together with the operating room staff. Furthermore, a high maneuverability leads to higher surgical safety, since it means that the manipulator can be easily and quickly removed from the surgical bed. Finally, small overall dimensions leave more space for the surgical team, for the nurse, and for the anesthesiologist's access to the patients, and thus, further improve overall surgical safety.
- *Improved operator's autonomy:* An optimal manipulator should be conceived with the main purpose to improve the surgical autonomy of the operator surgeon, allowing him/her to operate without the help of a skilled laparoscopic assistance. In this contest, the expertise level of the assistant surgeon should not represent a critical aspect of the procedure, unlike the current robotic system. Furthermore, a manipulator conceived to be small in dimensions and user-friendly could lead to an improvement in the ergonomics not only of the operating surgeon but also of the assistants.

Other more technical desired features that could increase the usability of the proposed robotic manipulator are as follows:

- *Arms' motion range:* The currently used robotic system arms (i.e., the da Vinci system) have external articulations that limit the motion range of the instruments inside the abdomen: since the internal articulation is limited to the end-effector EndoWrist, the system needs to move the arms outside the abdomen to change surgical target, increasing the fulcrum effect on the abdominal wall port-sites. Consequently, the robotic system is greatly related to the trocars' position, leading to a critical importance of the port-site position. Furthermore, robotic arms, with their external articulation, lead to difficulties in changing the surgical target inside the abdomen, and when it is needed to work in different anatomical districts, it is often necessary to move the entire robotic trolley. This can really limit the range of surgical procedures suitable for robotic surgery. In this context, a robotic arm potentially able to move itself inside the abdomen could greatly overcome these limitations, allowing more complex movements inside the abdomen and proportionally lower movements outside it.

- *Haptic feedback:* An arm which allows the surgeon to have a haptic feedback of the handled tissues is of critical importance in improving surgical performance and in avoiding tissue tears with potentially harmful consequences, such as bowel perforations.
- *Range of instruments:* From a clinical point of view, the robotic arm could not be conceived without envisioning the end-effector of the arm, i.e., the range of available instruments that can be used with the arm defines to a large extent the usability of the entire arm. A wide and complete range of instruments available for the robotic arm would allow avoiding the use of extra ports and the help of one or more assistants. Also, removing the robotic arm from the abdomen and changing the instrument should be easily and quickly done, in order to avoid a lengthening of operative time and a decrease in surgical safety when a rapid change is needed.

1.1.2 Technical Specifications

The characteristics qualitatively described in the previous section have to be translated into technical specifications to lead and steer the design and the fabrication of the device. It should show squeezing capabilities to be able to pass through a traditional trocar port or an umbilical access. An active, flexible, and articulated tip would improve the dexterity of the device thus allowing the performance of surgical tasks, such as suturing, cauterization, etc. The final device should be able to move through a bend of up to 270° around a large organ, grasp that organ, and retain a grip while moving through 20 mm. The manipulator main features consist of bending capability in any direction, active elongation, and selective stiffening.

Based on the above considerations, the specifications for the flexible arm can be summarized as follows:

- Capability to squeeze and pass through a 20 mm port; hence, the manipulator can be employed for umbilical single-port surgery, NOTES surgery [1], and single access surgery;
- Flexible and articulated length of up to 300 mm, enabling it to turn around organs in the thoracic and abdominal cavities, independently on the entrance point;
- Possible elongation of up to 100 mm ($\sim$33%);

- Force of at least 1 N (better if 5 N) to be achieved in stiffened condition at relevant points along the arm, including the tip, in order to meet typical force requirements of manipulation in surgical tasks like displacement of organs during arm motion or organ retraction [2, 3];
- The stiffness of the modules should be controllable, thus allowing to smoothly adapt to soft organs' geometries as well as to become rigid for retraction actions;
- A tool (such as a gripper) should be attached to the module tip and it should be able to exert a force up to 10 N.

Apart from the requirements listed above, haptic feedback and the possibility to combine the manipulator with a series of instruments need to be considered.

1.2 Technological Overview of Different Actuation Strategies

In this section, an overview of different actuation strategies is reported. This represents the very first step of the design phase where, given the manipulator specifications, the most suitable technologies are revised and compared. It is split into two main groups that refer to two different active capabilities: the first is "active motion" and it contains all the technological solutions that can be used to produce a force and a deformation that actively interact with the environment; the second one is "stiffness variation" and reports several solutions to actively vary the stiffness of structures.

1.2.1 Active Motion Technology Survey

The technological solutions considered for the realization of the active motion of the STIFF-FLOP manipulator are reported in this section. A streamlined but complete survey enables a direct comparison that eases the choice of the most suitable technology. This is followed by a detailed analysis of its state of the art.

Actuators represent the real bottleneck in many robotic applications and even if currently the most used technologies are electromagnetic-driven, they present limitations in terms of inertia and back-drivability, stiffness control, and power consumption.

1.2.1.1 Electromagnetic motors

Electromagnetic motors (cable-driven mechanisms or geared) may lead to several advantages in terms of controllability, actuation forces, and speed. Indeed, most of robotic devices use this technology for the actuation. Typically, the motion generated by electromagnetic motors is converted by an appropriate mechanism in order to generate the desired motion such as geared boxes, cables, or hybrid solutions. One of the main disadvantages of the electromagnetic actuators is that they are rigid and thus limit the flexibility of an instrument if embedded on-board. In addition, miniaturized motors do not have the required performances especially in terms of provided torque.

Cable actuation can be an alternative solution, since powerful motors can be embedded outside the robot thus keeping it flexible. The disadvantages of this solution are mainly in the friction losses along the robot due to the cables that may reduce the controllability of the system; in addition in order to provide an effective actuation at a high dexterity (hyper-redundancy), a high number of powerful and thus big actuators are needed externally (as in the case of the da Vinci, where the encumbrance in the surgery room is one of the main limiting factors).

1.2.1.2 Electro active polymers

In the last few years, new and promising technologies [4] are emerging thus offering new possibilities to fill the gap between natural muscles and artificial artifacts. Most of them are based on polymeric matrixes activated with different mechanisms (**Electro Active Polymers**, EAP) [5, 6].

Dielectric elastomers can be used with different structural configurations to perform an electrical squeezing [7]. The most interesting and functional solution consists in stacking many units composed by dielectric material and electrodes in order to exploit the axial contraction of each unit [8, 9]. *Ferroelectric polymers* have the characteristic to lose their natural polarization when over their Curie point. Zhang et al. found that the copolymers under a proper irradiation treatment exhibit very little room temperature polarization hysteresis and larger electrostrictive strain [10]. *Liquid crystal elastomers* (LCEs) are thermally actuated to produce macroscopic and anisotropic shape changes [11]. Stacks of very thin layers can be manufactured to produce fast and relatively high contractions [12]. *Conductive polymers* (or conjugated polymers) can change their dimensions thanks to the ability to lose or acquire ions when a voltage is applied [13] and most used materials are polypyrrole [14] and polyaniline. By using these materials, the authors of [15] produced PAN fibers and investigated the possibility to use them as linear actuators in wet and dry conditions. *Polymer gels* are able to swell through the uptake of a solvent within a polymer matrix and this process can be affected by electrical means. Gel actuators can exhibit relatively rapid response (0.1 s or less) provided that their surface area to volume ratios are high enough to reduce solvent diffusion times. At the University of Reading, researchers developed essentially Mckibben-type actuators in which the gel replaces the gas [16]. *Carbon nanotubes* were shown to generate higher stresses than natural muscle but lower strains than other polymer-based technologies [17]. *Ionic polymer/metal composites* [18] are based on migration of ions due to application of voltage. They are mainly used for the ability to bend themselves [19–22] but also to vary the stiffness of a structure [23].

These technologies show different performances, some advantages, and several constraints [24], but currently there are no satisfying solutions to imitate natural muscles [25].

1.2.1.3 Shape memory alloys

Shape memory alloys (SMA) [26] are increasingly used in biomimetic robots as bioinspired actuators. There are several kinds of alloys that can be used with different performances. The response time depends on the time needed to pass from the martensitic to the austenitic phase, and consequently it depends on the current that passes through the wires, the dimensions, and the thermal coefficient. A trade-off between current and velocity has to be found, but an acceptable frequency of contraction/relaxation can be reached with a well-designed geometry. SMA has a high force to mass ratio, is lightweight, and compactness, and, for these reasons, it represents a very interesting technology in the prosthetic field [27]. Nevertheless, SMA shows several limitations, like: difficulty in controlling the length of the fibers as they undergo the phase transition first of all due to their hysteresis; dependence of the bandwidth on heating and cooling rates; and limited lifecycles. These data depend on the percentage of contraction that has to be reached (300 for 5%, 10,000 for 0.5%) [24]. Despite that, there are several examples (especially in the biomimetic field) that demonstrate the real possibility to use this technology for actuation. Ayers, for example, developed a myomorphic actuator for his robotic lobster exploiting SMA wires and using pulse width duty cycle modulation to grade the proportion of martensite that transforms to austenite [28].

1.2.1.4 Shape memory polymers

Shape memory polymers (SMPs) belong to a class of smart polymers, which have drawn considerable research interest in the last few years because of their applications in micro electromechanical systems, actuators, for self-healing and health monitoring purposes, and in biomedical devices. Shape memory polymers exploit the same idea as SMAs: the configuration at the point of cross-linking in a polymer network is the lowest energy configuration. Any deformation away from that configuration increases the energy. Thus, by stretching the rubber at a temperature above the crystallization point and then cooling it in that extended shape will lock-in the deformed material. On heating to melt the crystals, the original network configuration is recovered. A variety of materials have been employed. Like in other fields of applications, SMP materials have been proved to be suitable substitutes to metallic ones because of their flexibility, biocompatibility, and wide scope of modifications. The shape memory properties of SMPs polymers might surpass those of shape memory metallic alloys (SMAs). A comprehensive review can be found in [29].

1.2.1.5 Flexible fluidic actuator

Flexible fluidic actuator is a term for a wide range of system types, but generically flexible fluidic actuators comprise an expansion chamber defined by an inner wall of an expandable girdle, the expandable girdle being connected to at least two anchoring points. The expansion chamber has at least one fluid inlet to allow pressurized fluid into the chamber. The expansion chamber is capable of acquiring a minimum volume and a maximum volume. Thus, these flexible actuators are able to adapt and transform a fluid pressure force against the inner wall of the expandable girdle and so produce a traction force or a bending movement among the two anchoring points, when the expansion chamber is inflated by the pressurized fluid entering through the fluid inlet.

1.2.2 Discussion and Choice of Active Motion Technology

The fluidic actuation presents nice features regarding an application inside the human body [30] (see table 1.1). Indeed, it has the non-negligible advantage to prevent having energized parts, i.e., being under electrical voltage (unlike electrostatic actuators, piezoelectric actuators [31], electroactive polymers, or electromagnetic motors when used inside the body), or high-temperature parts (unlike the SMA and thermal actuators) inside a patient's body; this increases safety. As no electrical power is used, operation in the presence of radioactivity or magnetic fields is possible [32]. In the case of a hydraulic

Table 1.1 Comparison table of several candidate technologies for the active motion of the module: red–unacceptable; light red–undesired; light green–acceptable; and green–desired

	input energy domain	scaling of dimensions	hard/soft type	Strain (%)	stress max (MPa)	power (W/cm^3)	Response time	Bandwidth (Hz)	Max strain rate (%/s)	Efficiency (%)
motor-driven	electric	limited	hard	0.5	0.1	0.1	fast	20	>1000	>80
SMA driven	electric and thermal	high	soft	0.1	200	0.35	medium-slow	3	100	3
Electroactive polymers	electric and thermal	high	soft	0.5	0.3	0.75	medium-slow	10	NA	30
Piezoelectric actuation	electric	high	hard	0.002	35	175	fast	5000	>1000	50
magnetostrictive actuators	magnetic	medium-high	soft	0.002	35	70	medium-slow	2000	NA	80
Electrostrictive polymer actuated muscles	electric	medium-high	soft	0.5	0.3	0.75	medium-fast	10	NA	30
Hydraulic actuators based	Fluid	medium-high	medium soft	0.5	20	20	medium-fast	4	100	80
Pneumatic actuators based	gas	medium-high	soft	0.5	0.7	3.5	fast	20	70	90

actuation, a sterile physiological saline solution could be used so that a leakage of the system would have no consequence on the patient's body.

One can think of miniaturizing classical piston-based fluidic actuators, but it raises difficulties regarding the sealing of the chambers. O-rings and lip seals are no longer suitable because small variations of the shape or size of the components (seal, seal house, or piston) involve high friction or leakage. De Volder et al. [33] propose to use "restriction seals," i.e., small clearances between the rod and the orifice. These generate less friction and allow a compromise between the leakage and the manufacturing accuracy; the actuator can present virtually no leakage, but then tolerances in the range of 1 μm or less are required. However, to avoid leakages and friction, which limit efficiency, pressurized elastic deformable chambers are preferred to be used, i.e., flexible fluidic actuators, as suggested by Suzumori et al. [34].

As these actuators present no relative motion of parts, static sealing can be used and this means no need for lubricants, no leakages, and no wear particles; consequently, these actuators could possibly operate in clean room, food, or agriculture industries [35]. Besides, smooth motion and precise positioning are possible to achieve since there is no friction [34] (unlike piston-based actuators or systems actuated with cables). In the field of robotics, compliant structures have relevant additional.

Advantages over traditional rigid body robots:

1. They can handle delicate objects without causing any damage thanks to their inherent compliance [35]. This compliance allows them to adapt themselves to their environment during contacts [35, 36].
2. Compared to traditional mechanisms made of articulated rigid parts, compliant structures allow the reduction of the number of parts necessary to perform a given task [37]. This is an interesting feature regarding miniaturization.
3. When they are made of membranes, flexible structures can be very lightweight. If the instrument is actuated thanks to inflatable membranes, its volume may be reduced when the membranes are deflated. This is an interesting characteristic if the whole device has to be inserted into a small orifice.

The combination of a fluidic actuation and a flexible structure also brings **advantageous properties**:

1. Regarding a medical application, reducing the fluid pressure, the device loses its rigidity and recovers its initial shape. In emergency cases, it

offers the opportunity to take the instrument out of the patient's body quickly.

2. Devices based on this principle and whose actuation is obtained by the deformation of elastic chambers have the intrinsic and very useful property of giving to the operator a reliable feedback about their posture and actuation force by measuring the volume and pressure of the operating fluid that has been supplied.

Nevertheless, fluidic actuation comes with some **drawbacks**:

1. Fluidic actuation needs equipment such as pumps, valves, and pipes that can be bulky. However, in the case of a medical application, the pump is placed outside the patient's body and will not increase the bulkiness of the instrument inside the body.
2. Regarding fluidic micro-actuators: the pipes used to drive the fluid can present leakages and cause pressure losses that limit efficiency [38]. Moreover, controlling fluid pressures and flow rates in small sections is often more delicate than controlling electrical quantities.
3. Another shortcoming of flexible fluidic actuators is due to the required control as explained in [35]: Fluidic flexible robots require sophisticated controllers in order to reach accurate and repeatable positioning. Moreover, the deformable structure and non-conventional actuation make their dynamics modeling very complex.

Finally, regarding the fluids used to inflate these actuators, one has characteristics that compensate the lacks of the other and vice versa. Liquids can generate higher forces and, in general, their supply circuit is safer since it has reduced possibilities of explosions when compared with gases. In addition, the compressibility of gases brings about more compliance, so that it leads to a more difficult characterization and modeling since it also involves thermal losses upon compression. On the other hand, air is a readily available source and exhaust gases (i.e., air) can be freely evacuated in the environment [39]. Furthermore, gases lead to more lightweight actuators and to pressure losses 1,000 times smaller than those for liquids [40].

1.2.3 Stiffness Variation Technology Survey

The technological solutions considered for the realization of the stiffness variation of the STIFF-FLOP manipulator are reported in this section. The survey is followed by a comparison table that guides the choice of the most suitable technology.

Stiffness variation is one of the main features of the STIFF-FLOP arm. The arm should be able to safely interact with the surgical environment adapting its stiffness according to the situation and the surgical task. Indeed the arm should be able to navigate in the body cavities in a floppy state and then selectively stiffen some of its segments to actively move organs or accomplish surgical tasks. Generally, navigation in the body cavities is performed using flexible medical instruments such as conventional endoscopes. These instruments are used because of their high flexibility, which enables traversing tortuous trajectories and the reaching of many different surgical targets, possibly even without the need to make skin incisions. Novel surgical instrumentation is being developed in order to exploit the higher dexterity, flexibility, and potential for miniaturization of these instruments. Many prototypes for robotic surgery and endoscopy have been developed and commercialized [41–43]. Typically such flexible instruments are composed by a flexible "backbone" or springs [44] and are actuated by motors located externally. The flexible backbone causes them to have low stiffness and makes it difficult to control the rigidity [45].

The stiffness variation in endoscopic instruments has been widely investigated [46]. Various stiffening strategies have been developed and have been considered in the choice of the STIFF-FLOP arm stiffening mechanism. Following the approach described in [46], both rigidity controls based on material stiffening and structural stiffening have been considered and reviewed. Rigidity control based on material stiffening exploits the change in the mechanical properties on particular materials due to controlled physical stimuli. Examples are phase change of **thermoplastic polymers** [47–53]. Phase change induced by temperature change can be used to change the stiffness of thermoplastic polymers from values resembling low viscosity fluids to values resembling rigid nylon. The main drawbacks of phase change polymers are that they are difficult to control and have low response time slow (order of second) since they rely on heating and cooling systems.

Other materials that can be used for stiffening varying their mechanical properties are electro and magneto rheological fluids. Electrorheological fluids change their viscosity in response to an electric field. They have been proposed for different applications [54, 55]. However, this kind of fluid requires high electric field, for example, in [56] and [57], it is reported that in this case, an electric field up to 5,000 V/mm at 2–15 mA/cm^2 is required to obtain yielding strength change from about 0 to 5 kPa, turning from liquid to quasi-solid in few milliseconds. Magnetorheological fluids work in a similar way to the electrorheological ones but they respond to a magnetic field. Although they are more energy efficient than electrorheological fluids [56, 57], they need high magnetic field (239 kA/m) for rigidifying a device (maximum yielding strength of 100 kPa). Such field would require highly encumbering magnetic sources.

Rigidity control based on material stiffening can be useful for precise control of damping and is mostly used in active damping mechanisms (tunable automotive suspensions), not for drastic changes in elastic modulus as required for robotic applications. As described above, sole material stiffening would not fit the stiffening requirements of the STIFF-FLOP arm.

Stiffening of a flexible structure can be obtained by locking the relative movements between interconnected parts of a structure (structural stiffening). As shown in Table 1.2, structural stiffening can act both on the angle of each successive segment of a flexible backbone (angle locking) and on the lengths of the inner and outer curves of the bends in a shaft-guide that are locked (curve length locking).

Both angle locking and curve length locking can be discrete or continuum. The discrete angle is mainly based on the increase in friction between two consequent joints due to the tensioning of the structure by means of cables or other tensioning systems. Applications can be found in [58–65].

In the discrete curve length locking strategy, the distance between the outer edges of succeeding elements is fixed [66] by means of cables, fluid columns, rods, or any other element that can lock and unlock the distance between two points.

Both discrete curve length locking and angle locking mechanisms require relatively large amounts of mechanical components and therefore are not simple to produce or scale down. In addition, since they mainly rely on friction between two or more components, stiffening control in not much controllable; indeed, they are mainly used as on–off mechanisms. In the case of continuum structural stiffening, the stiffness variation occurs continuously along a structure, not only between two or more segments. This kind of stiffening can be used both for angle locking and for curve length locking. Recently, Loeve et al. [67] presented a continuum structural stiffening strategy based on friction between a central fluidic channel and the cable actuation mechanism. Otherwise, it can be implemented by exploiting vacuum packed particles [68–74].

This strategy is based on the physical phenomenon of **granular jamming**. Granular jamming is a growing field in robotics, and is a mechanism which enables particles to act like a liquid, solid, or something in between depending on an applied pressure. As stated by [75], jamming is a phenomenon where an external stress can change "fragile matter" from a fluid-like to a solid-like state. Because of this unique feature, many groups have integrated granular jamming into robotic projects such as the universal robotic gripper [76], the tendon-supported elephant trunk, the jamming skin-enabled locomotion robot [77, 78], the variable stiffness haptic device [79], the variable stiffness endoscope [46], and the emergency vacuum splint [80].

1.2.4 Comparison and Choice

Granule or particle jamming has interesting features such as high deformability in the fluid state, and drastic stiffness increase in the solid state, without significant change in volume; in addition, it allows controlling the stiffness level by controlling the level of an applied vacuum (see table 1.2). Due to these unique features, it is currently the design choice explored for the stiffening mechanism of the STIFF-FLOP arm. Indeed, it is the most suitable solution for highly deformable soft structures.

Table 1.2   Comparison table of several candidate technologies for the variation of the module stiffness: red–unacceptable; light red–undesired; light green–acceptable; and green–desired

Stiffening Strategy		Physical phenomenon involved	Controllability	Responce time	Stiffening range
Material stiffening	Phase change polymers	phase change of thermoplastic polymers	Low	order of second	From low viscosity fluids to values resembling rigid nylon
	Magnetorehology	changes their viscosity in response to magnetic field	Low (difficoult to tune stiffness)	millisecond	maximum yielding strength of 100kPa (239 kA/m magnetic field)
	Electrorehology	changes their viscosity in response to electric field	Low (difficoult to tune stiffness)	millisecond	yielding strength change from about 0 to 5 kPa (5,000 V/mm at 2–15 mA/cm²)
Structural stiffening	Descrete Angle locking	Friction between two consequent joints due to the tensioning of the structure	Low, mainly on-off	High	Shapelocking capabilities applying high tensioning forces
	Descrete Curve lenght locking	the distance between the outer edges of succeeding elements is fixed by friction	Low, mainly on-off	High	Shapelocking capabilities applying high tensioning forces
	Continous angle locking and curve lenght locking	This strategy is based on the physical phenomenon of the granular jamming.	Possible controlling the vacuum level	High, dependent on performances of the vacuuming system	high deformability in the fluid state, and drastic stiffness increase in the solid state

References

[1] Auyang, E., Santos, B., Enter, D., Hungness, E., and Soper, N. (2011). Natural orifice translumenal endoscopic surgery: a technical review. *Surg. Endosc.* 25, 3135–3148.

[2] Brown, J. D., Rosen, J., Chang, L., Sinanan, M. N., and Hannaford, B. (2004). Quantifying surgeon grasping mechanics in laparoscopy using the Blue DRAGON system. *Med. Meets Virtual Reality* 13, 34–36.

[3] Richards, C., Rosen, J., Hannaford, B., Pellegrini, C., and Sinanan, M. (2000). Skills evaluation in minimally invasive surgery using force/torque signatures. *Surg. Endosc.* 14, 791–798.

[4] Pons, J. (2005). *Emerging Actuator Technologies: A Micromechatronic Approach.* Chichester: John Wiley & Sons Ltd.

[5] Bar-Cohen, Y. (2004). *Electroactive Polymer (EAP) Actuators as Artificial Muscles: Reality, Potential, and Challenges.* Bellingham, WA: SPIE Press.

[6] Kratz, R., Stelzer, M., Friedmann, M., and von Stryk, O. (2007). "Control approach for a novel high power-to-weight ratio SMA muscle scalable in force and length," in *Proceedings of the IEEE/ASME International Conference on Advanced Intelligent Mechatronics (AIM)*, (Zurich, CH).

[7] Pelrine, R. (2002). "Dielectric elastomer artificial muscle actuators: towards biomimetic motion," in *Proceedings of the SPIE Smart Structures and Materials, Electroactive Polymer Actuators and Devices*, 4695, 126–137.

[8] Carpi, F., Salaris, C., and Rossi, D. D. (2007). Folded dielectric elastomer actuators. *Smart Mater. Struct.* 16, S300–S305.

[9] Laschi, C., Mazzolai, B., Cianchetti, M., Mattoli, V., Luciani, L. B., and Dario, P. (2008). "Design of a biomimetic robotic octopus arm," in *Proceedings of the Biological Approaches for Engineering Conference Proceedings*, 54–57.

[10] Cheng, Z. Y., Bharti, V., Xu, T. B., Xu, H., Mai, T., and Zhang, Q. M. (2001). Electrostrictive poly(vinylidene fluoride-trifluoroethylene) copolymers. *Sens. Actuat. A: Phys.* 90, 138–147.

[11] Lehmann, W., Hartmann, L., Kremer, F., Stein, P., Finkelmann, H., Kruth, H., and Diele, S. (1999). Direct and inverse electromechanical effect in ferroelectric liquid crystalline elastomers. *J. Appl. Phys.* 86, 1647–1652.

[12] Spillmann, C., Naciri, J., Martin, B., Farahat, W., Herr, H., and Ratna, B. (2006). Stacking nematic elastomers for artificial muscle applications. *Sens. Actuat. A* 133, 500–505.

[13] Baughman, R. (1996). Conducting polymer artificial muscles. *Synth. Metals.* 78, 339–353.

[14] Ryu, J., Park, J., Kim, B., and Park, J.-O. (2005). Design and fabrication of a largely deformable sensorized polymer actuator. *Biosens. Bioelectr.* 21, 822–826.

[15] Mazzoldi, A., Degl'Innocenti, C., Michelucci, M., and Rossi, D. D. (1998). Actuative properties of polyaniline fibers under electrochemical stimulation. *Mater. Sci. Eng. C.* 6, 65–72.

[16] Santulli, C., Patel, S. I., Jeronimidis, G., Davis, F. J., and Mitchell, G. R. (2005). Development of smart variable stiffness actuators using polymer hydrogels. *Smart Mater. Struct.* 14, 434.

[17] Baughman, R. H., Cui, C., Zakhidov, A. A., Iqbal, Z., Barisci, J. N., Spinks, G. M., et al. (1999). Carbon nanotube actuators. *Science* 284, 1340–1344.

[18] Kim, K., and Shahinpoor, M. (2002). Development of three dimensional ionic polymer-metal composites as artificial muscles. *Polymer* 43, 797–802.

[19] Zhang, S. W., Guo, S., and Asaka, K. (2006). "Characteristics analysis of a biomimetic underwater walking microrobot," in *Proceedings of the IEEE International Conference on Robotics and Biomimetics, 2006 (ROBIO '06)*, 1600–1605.

[20] Wang, X.-L., Oh, I.-K., Lu, J., Ju, J. and Lee, S. (2007). Biomimetic electro-active polymer based on sulfonated poly (styrene-*b*-ethylene-*co*-butylene-*b*-styrene). *Mater. Lett.* 61, 5117–5120.

[21] Shahinpoor, M. (2003). Ionic polymer-conductor composites as biomimetic sensors, robotic actuators and artificial muscles–a review. *Electrochimica Acta* 48, 2343–2353.

[22] Sugiyama, K., Ishii, K., and Kaneto, K. (2006). Development of an oscillating fin type actuator for underwater robots. *Int. Congr. Ser.* 1301, 214–217, 2007. Brain-Inspired IT III. *Invited and selected papers of the 3rd International Conference on Brain-Inspired Information Technology 2006*, Kitakyushu.

[23] Kobayashi, S., Ozaki, T., Nakabayashi, M., Morikawa, H., and Itoh, A. (2006). "Bioinspired aquatic propulsion mechanisms with real-time

variable apparent stiffness fins," in *Proceedings of the IEEE International Conference on Robotics and Biomimetics, 2006 (ROBIO'06)*, 463–467.

[24] Madden, J., Vandesteeg, N., Anquetil, P., Madden, P., Takshi, A., Pytel, R., et al. (2004). Artificial muscle technology: physical principles and naval prospects. *IEEE J. Oceanic Eng.* 29, 706–728.

[25] Mirfakhrai, T., Madden, J. D., and Baughman, R. H. (2007). Polymer artificial muscles. *Mater. Today* 10, 30–38.

[26] Funakubo, J. B., and Kennedy, H. (1987). *Shape Memory Alloys*. Philadelphia, PA: Gordon and Breach Science Publishers.

[27] Bundhoo, V., and Park, E. (2005). "Design of an artificial muscle actuated finger towards biomimetic prosthetic hands," in *Proceedings of the 12th International Conference on Advanced Robotics, 2005 (ICAR '05)*, 368–375.

[28] Ayers, J. (2004). Underwater walking. *Arthropod Struct. Develop.* 33, 347–360.

[29] Ratna, D., and Karger-Kocsis, J. (2008). Recent advances in shape memory polymers and composites: a review. *J. Sc. Mater.* 43, 254–269.

[30] De Greef, A., Lambert, P., and Delchambre, A. (2009). Towards flexible medical instruments: review of flexible fluidic actuators. *Precis. Eng.* 33, 311–321.

[31] Thomann, G. (2003). *Contribution á la Chirurgie Minimalement Invasive: Conception d'un Coloscope Intelligent*. Ph.D. thesis, National Institute of Applied Sciences, Lyon.

[32] Bertetto, A. M., and Ruggiu, M. (2004). A novel fluidic bellows manipulator. *J. Robot. Mechatr.* 16, 604–612.

[33] De Volder, M., Peirs, J., Reynaerts, D., Coosemans, J., Puers, R., Smal, O., et al. (2005). Production and characterization of a hydraulic microactuator. *J. Micromech. Microeng.* 15, 15–21.

[34] Suzumori, K., Matsumaru, T., and Likura, S. (1990). *Elastically Deformable Fluid Actuator*. U.S. Patent 4,976,191.

[35] Bertetto, A. M., and Ruggiu, M. (2004). A novel fluidic bellows manipulator. *J. Robotics Mechatr.* 16, 604–612.

[36] Prelle, C. (1997). *Contribution au Contrôle de la Compliance d'un Bras de Robot á Actionnement Électropneumatique*. Ph.D. thesis, Institut National des Sciences Appliquées, Lyon.

[37] Howell, L. L. (2001). *Compliant Mechanisms*. Hoboken, NJ: John Wiley & Sons.

[38] Anthierens, C. (1999). *Conception d'un Micro Robot á Action Neurasservi Électropneumatique pour l'Inspectionintratubulaire.* Ph.D. thesis, Institut National des Sciences Appliquées, Lyon.

[39] Smith, S. T., and Seugling, R. M. (2006). Sensor and actuator considerations for precision, small machines. *Precis. Eng.* 30, 245–264.

[40] Taillard, P. (2000). Guide de dimensionnement: La production d'énergiepneumatique. *Technologie* 110, 21–25.

[41] Ding, J., Xu, K., Goldman, R., Allen, P., Fowler, D., and Simaan, N. (2010). "Design, simulation and evaluation of kinematic alternatives for insertable robotic effectors platforms in single port access surgery," in *Proceedings of the IEEE International Conference on Robotics and Automation—ICRA*, 1053–1058.

[42] Thompson, C., Ryou, M., Soper, N., Hungess, E., Rothstein, R., and Swanstrom, L. (2009). Evaluation of a manually driven, multitasking platform for complex endoluminal and natural orifice transluminal endoscopic surgery applications. *Gastrointest. Endos.* 70, 121–125.

[43] Degani, A., Choset, H., Zubiate, B., Ota, T., and Zenati, M. (2008). "Highly articulated robotic probe for minimally invasive surgery," in *Proceedings of the IEEE International Conference EMBS*, Vancouver, BC, 3273–3276.

[44] Gravagne, I., and Walker, I., (2002). "Manipulability, force, and compliance analysis for planar continuum manipulators," in *Proceedings of the IEEE Transactions on Robotics and Automation*, 18, 3.

[45] Kim, Y.-J., Cheng, S., Kim, S., and Lagnemma, K. (2012). "Design of a tubular snake-like manipulator with stiffening capability by layer jamming," in *Proceedings of the IEEE/RSJ International Conference on Intelligent Robots and Systems (IROS), 2012 (IROS 2012)*, Vilamoura.

[46] Loeve, A., Breedveld, P., and Dankelman, J. (2010). "Scopes too flexible... and too stiff," in *Proceedings of the IEEE Pulse*, 1, 26–41.

[47] Sokolowski, W. (2004). *Cold Hibernated Elastic Memory Self-deployable and Rigidizable Structure and Method Therefor.* U.S. Patent 6,702,976.

[48] Griffin, S. (2006). *Selectively Flexible Catheter and Method of Use.* WO Patent 2006/060312.

[49] Guidanean, K., and Lichodziejewski, D. (2002). "An inflatable rigidizable truss structure based on new sub-TgPlyurethane composites," in *Proceedings of the AIAA/ASME/ASCE/AHS/ASC Structures, Structural Dynamics, and Materials Conference.* Denver, CO.

[50] Eidenschink, T. (2005). *Variable Manipulative Strength Catheter*. WO Patent 2005/016430.

[51] Redell, F. H., Lichodziejewski, D., Kleber, J., and Greschik, G. (2005). "Testing of an inflation-deployed sub-Tg rigidized support structure for a planar membrane waveguide antenna," in *Proceedings of the Collection of Technical Papers—AIAA/ASME/ASCE/AHS/ASC Structures, Structural Dynamics and Materials Conference*, Austin, TX, 920–927.

[52] Cheng, N., Ishigami, G., Hawthorne, S., Hao, C., Hansen, M., Telleria, M., et al. (2010). "Design and analysis of a soft mobile robot composed of multiple thermally activated joints driven by a single actuator," in *Proceedings of the IEEE International Conference on Robotics and Automation—ICRA*, Anchorage, CA, 5207–5212.

[53] Telleria, M., Hansen, M., Campbell, D., Servi, A., and Culpepper, M. (2010). "Modeling and implementation of solder-activated joints for single-actuator, centimeter-scale robotic mechanisms," in *Proceedings of the IEEE International Conference on Robotics and Automation—ICRA*, Anchorage, CA, 1681–1686.

[54] Olson, G. (2007). *Medical Devices with Variable Stiffness*. WO Patent 2007/015981.

[55] Rudloff, D. A. C. (1987). *Penile Implant*. U.S. Patent 4,664,100.

[56] Park, G., Bement, M. T., Hartman, D. A., Smith, R. E., and Farrar, C. R. (2007). The use of active materials for machining processes: A review. *Int. J. Mach. Tools Manuf.* 47, 2189–2206.

[57] Yalcintas, M., and Dai, H. (1999). Magnetorheological and electrorheological materials in adaptive structures and their performance comparison. *Smart Mater. Struct.* 8, 560–573.

[58] Saadat, V., Rothe, C. A., Ewers, R. C., Maahs, T. D., and Michlitsch, K. J. (2006). *Endoluminal Tool Deployment System*. U.S. Patent 2006/0178560.

[59] Tartaglia, J. M., Keller, W. A., Belson, A., and Ratchford, A. R. (2006). *Endoscope with Adjacently Positioned Guiding Apparatus*. U.S. Patent 6,984,203.

[60] Zehel, W. E., Baumann, D. M., and Brenner, W. B. (1993). *Method and Apparatus for Conducting Exploratory Procedures*. U.S. Patent 5,251,611.

[61] Sturges, R. H. J. R., and Laowattana, S. (1993). A flexible, tendon-controlled device for endoscopy. *Int. J. Robot. Res.* 12, 121–131.

[62] Vargas, J. S. (2006). *Cannula System and Method of Use*. U.S. Patent 2006/0025652.

[63] Timm, G. W., Helms, R. A., and Sandford, D. L. (1985). *Penile Prosthesis Utilizing a Patient Controlled Camactuator Apparatus*. WO Patent 85/04797.

[64] Saadat, V., Rothe, C. A., Ewers, R. C., Maahs, T. D., and Michlitsch, K. J. (2006). *Endoluminal Tool Deployment System*. U.S. Patent 2006/0178560.

[65] Brenneman, R., Ewers, R. C., Saadat, V., and Chen, E. G. (2004). *Apparatus and Methods for Guiding Anendoscope via a Rigidizable Wire Guide*. U.S. Patent 2004/0186350.

[66] Yagi, A., Matsumiya, K., Masamune, K., Liao, H., and Dohi, T. (2006). "Rigid-flexible outer sheath model using slider linkage locking mechanism and air pressure for endoscopic surgery," in *Proceedings of the Medical Image Computing and Computer-Assisted Intervention (MICCAI '06)*, 503–510.

[67] Loeve, A. J., Plettenburg, D. H., Breedveld, P., and Dankelman, J. (2012). "Endoscope shaft-rigidity control mechanism: "FORGUIDE," in *IEEE Transactions on Biomedical Engineering*, 59, 542–551.

[68] Zinner, N. R., and Sterling, A. M. (1991). *Penile Prosthesis and Method*. U.S. Patent 5,069,201.

[69] Campanaro, L., Goldstone, N. J., and Shepherd, C. C. (1966). *Rigidized Evacuated Structure*. U.S. Patent 3,258,883.

[70] Rose, F. L. (1973). *Vacuum Formed Support Structures and Immobilizer Devices*. U.S. Patent 3,745,998.

[71] Loeb, J., and Plantif, B. E. P. J. (1978). *Systeme de Protection par Modelage sous Formed'enceintedeformable et Rigidifiable par Depression*. CA Patent (Brevet Canadien) 1035055.

[72] Loeve, A. J., van de Ven, O. S., Vogel, J. G., Breedveld, P., and Dankelman, J. (2010). Vacuum packed particles as flexible edoscope guides with controllable rigidity. *Gran. Matt.* 12, 543–554.

[73] Marcus, D. (2002). *Controlled Rigidity Articles*. WO Patent 02/002309.

[74] Barrington, J. (1979). *Implantable Penile Prosthesis*. U.S. Patent 4,151,841.

[75] Liu, A. J., and Nagel, S. R. (1998). Jamming is not just cool anymore. *Nature* 396, 21–22.

[76] Brown, E., Rodenberg, N., Amend, J., Mozeika, A., Steltz, E., Zakin, M. R., et al. (2010). "Universal robotic gripper based on the jamming of granular material," in *Proceedings of the National Academy of Sciences of the United States of America*, 107, 18,809–18,814.

[77] Steltz, E., Mozeika, A., Rodenberg, N., Brown, E., and Jaeger, H. (2009). "Jsel: Jamming skin enabled locomotion," in *Proceedings of the IEEE/RSJ International Conference on Intelligent Robots and Systems, 2009 (IROS 2009)*, 5672–5677.

[78] Steltz, E., Mozeika, A., Rembisz, J., Corson, N., and Jaeger, H. (2010). "Jamming as an enabling technology for soft robotics," in *Proceedings of the Society of Photo-Optical Instrumentation Engineers (SPIE) Conference Series*, 7642, 63.

[79] Mitsuda, T., Kuge, S., Wakabayashi, M., and Kawamura, S. (2002). "Wearable haptic display by the use of a particle mechanical constraint," in *Proceedings of the 10th Symposium on HAPTICS 2002*, 153–158.

[80] Letts, R., and Hobson, D. (1973). The vacuum splint: an aid in emergency splinting of fractures. *Can. Med. Assoc. J.* 109, 599–600.

2

Design of the Multi-module Manipulator

**Tommaso Ranzani, Iris de Falco, Matteo Cianchetti
and Arianna Menciassi**

The BioRobotics Institute, Scuola Superiore Sant'Anna,
Pontedera (PI), Italy

Abstract

In Chapter 1 all the suitable technologies have been surveyed and a direct comparison underlined how fluidic technologies have the best characteristics to meet the application scenario. In particular, flexible fluidic actuators have been identified as the most promising technology for providing omnidirectional bending and elongation, while granular jamming can represent a valid solution to implement variable stiffness features. In this chapter, we report the design of the single module, the strategy for integrating more modules, the fabrication and characterization of a 2-module manipulator.

2.1 The Design of the Single Module

Each module of the manipulator has to be able to independently perform omnidirectional bending, elongation, and stiffening. This is possible thanks to two different actuation systems integrated in each module: flexible fluidic actuators combined with a chamber exploiting the granular jamming mechanism.

The main component of the manipulator module is an elastomeric cylinder (silicone Ecoflex^{TM} 0050, Smooth-on Inc.). This material guarantees the right level of softness when deformed passively and it is suitable to host internal chambers that can be used to modulate the characteristics and the behavior of the module. The cylindrical elastomer hosts three equally spaced chambers which are embedded in radial arrangement (the fluidic actuators) and another one centrally placed (for granular jamming) as shown in

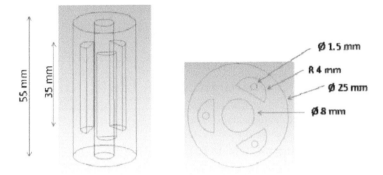

Figure 2.1   From left to right, sketch of the longitudinal and transversal cross section design of the module with the semicylindrical fluidic chambers and the central stiffening channel.

Figure 2.1. Externally the module is provided with a braided bellow-like structure.

2.1.1 Active Motion

Flexible fluidic actuators have been already successfully used as an active motion system enabling elongation and bending of soft structures [1]. The use of such technology is eased by the widely available literature in terms of modelization [2, 3] and application cases [4].

Optimal geometries for this specific system are under investigation, but previous works comparing several cross section designs for similar applications [5] concluded that the key factor is to find a trade-off between the thickness of the separation wall among the chambers and their diameter. Moreover, in this case an additional criterion is the maximization of the available internal space to host the stiffening chamber.

A well-known drawback on the use of soft material chambers inflated by pressurized fluids is the difficulty to have deformations along preferential directions to obtain bending and elongation. This is due to the fact that the inflated chambers tend to expand in every direction, like balloons. In Figure 2.2, the effect of 0.32 bar pressure on one chamber is shown and it is evident that the outward expansion is dominant with respect to the bending of the module and it reaches an unacceptably high risk of explosion.

On the contrary, in order to produce a more pronounced bending effect with minimal lateral expansion, the radial expansion of the chamber should

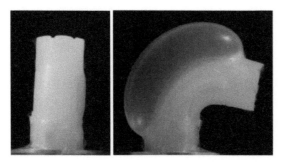

Figure 2.2 Effect of 0.32 bar pressure on a single silicone (Ecoflex 0050) chamber.

be minimized inducing a maximal longitudinal deformation. Since the elastomeric materials are considered isotropic, there are no preferential directions of expansion; however, adding other structures (fillers or external constrains) would force the elongation or the bending, limiting the diameter expansion and changing the overall module behavior.

Previous attempts to limit this lateral expansion demonstrated that circular fibers arranged all around the structure can serve the scope [6], but as the chamber deforms, the coils start to separate, leaving weaker areas on the external surface that could likely cause abrupt and dangerous lateral expansions. This risk is especially high if the elastomeric material is very soft and the achievable curvature is high. On the other hand, harder elastomers or a huge number of stiff fibers could compromise the performances of the manipulator or require a more powerful fluidic source.

Based on the above considerations, the main idea is to couple the silicone cylinder and its internal chambers with a braided structure (i.e., a sheath). Braided structures (like those used in the McKibben actuators) are highly flexible and can contemporarily follow bending and elongation movements providing a radial constraint to the excessive expansion especially if thermally formed to remain in a bellows-like shape.

The braided sheath is placed externally respect to the chambers and it is fixed at the distal ends of the cylinder. When the chambers are inflated to bend (or to elongate), the braided sheath contains and limits the radial expansion, thus maximizing the longitudinal effect of the deformation.

2.1.2 Stiffness variation

For stiffness modulation, a granular jamming solution is used. The effectiveness of this strategy on soft robots has been already demonstrated in [7–12].

One of the most interesting features of this technology is that it keeps a high deformability in the unjammed state and undergoes a drastic stiffness increase in the jammed condition. In our application, coffee powder was used as granular material and latex as containing membrane. Jamming transition is induced by increasing density in the flexible membrane due to the applied vacuum. By controlling the vacuum level the stiffness can be tuned.

2.2 Connection of Multiple Modules

The easiest strategy for module integration (in case of a limited number of modules) is based on the connection between the modules with pneumatic tubes that pass through the actuation chambers. This configuration allows aligning the chambers and having more free space within the module section.

A first prototype of the manipulator has been fabricated integrating three modules, designed on the basis of the approach reported in the previous section. The manipulator allowed evaluating the possible movements and optimizing the integration process for the fabrication of a second arm. This manipulator includes two modules and the geometry and dimensions of the single silicone unit are the same as that of the prototype illustrated in the previous section. The modules are connected with small silicone tubes passing through the chambers of the first module, as illustrated in Figure 2.3.

The connecting junction between two modules is the main element that has to be considered (length h in Figure 2.3). This area is not actuated and it does not include the granular jamming-based stiffening chamber. The junction connection has been designed in order to minimize its non-active effect on the system performance and not to prevail on that.

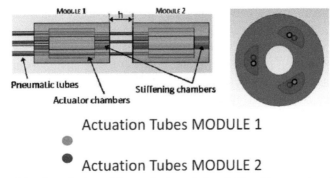

Figure 2.3   Overall view of the manipulator design (left) and bottom view (right).

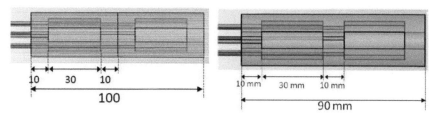

Figure 2.4 Theoretical length of the manipulator when two modules are connected (left) and manipulator with a junction of 10 mm (right).

If the manipulator is solicited laterally on the tip by a force, the designed configuration avoids a deflection of the manipulator in correspondence with the junction. In this way, it is guaranteed that, when evaluating the stiffening capabilities, the overall behavior of the manipulator will not be affected by the presence of this softer and non-active part. The theoretical length of the two-module manipulator is 100 mm which is twice the length of a single module as illustrated in Figure 2.4 (left).

The section between the modules has a length of 20 mm. However, 10 mm is sufficient for avoiding deformation of the silicone in the axial direction when it is inflated (see Figure 2.4, right panel). Therefore, a single junction of 10 mm has been designed between the chambers of the modules that will be functional for both the lower and upper modules.

Taking into account the geometry of the manipulator and the stiffness of the materials, an estimation of the junction displacement is given by:

$$y = \frac{Fh^3}{3EI} \tag{2.1}$$

where:

F is the applied force (N);

h is the junction length (m);

E is the elastic modulus of the junction (Pa);

I is the area moment of inertia (m^4) that is $\frac{\pi}{4}r^4$ for a filled circular area of radius r.

When a force is applied to the second module, the junction displacement can be controlled by dimensioning its length and material. The junction has been designed considering a small length of 8 mm and the material Dragon Skin 10 MEDIUM (Smooth-On) which is a silicone harder than Ecoflex 00-50 (Figure 2.5).

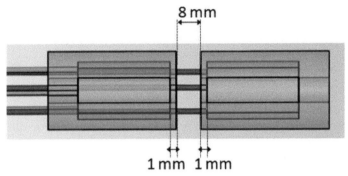

Figure 2.5   CAD of the junction zone.

For a force of 3 N, a length of 8 mm and a stiffness of 228 kPa (Dragon Skin), the displacement of the passive junction is:

$$y = \frac{3N \cdot (8mm)^3}{3 \cdot 228kPa \cdot \frac{\pi}{4} \cdot (12.5mm)^4} = 0.12mm \tag{2.2}$$

This displacement is negligible with respect to the displacement of the active unit, which is about 11 mm with a load force of 3 N as it is reported in Figure 2.6.

This configuration ensures a working length of 50 mm for each module and the effect of the junction is limited (see Figure 2.7).

In the 2-module manipulator the "in-series" stiffening mechanism has been inserted inside the central channel. Two membranes have been filled

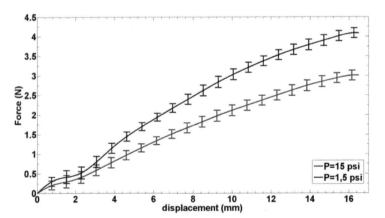

Figure 2.6   Force-displacement curve of the single module.

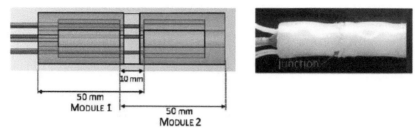

Figure 2.7 Final CAD of the 2-module manipulator (left) and the prototype (right).

Figure 2.8 Final prototype of the 2-module manipulator.

with 4 g of coffee grains and they have been connected with a tube of 4 mm diameter and 20 mm length. The vacuum has been applied simultaneously to the two membranes. Two external braided sheaths have been also integrated around the module. The final prototype is reported in Figure 2.8.

2.3 Complete Characterization of the 2-Module Manipulator

The soft manipulator is ideally composed of multiple modules, each one provided with actuation and stiffening capabilities. Starting from the design of the single module reported above and after the complete characterization of a single module reported in [13], we here extend the same analysis to a 2-module manipulator. This step is particularly significant since it allows—with a minimum number of modules—the testing of the combined performance of two interconnected modules in terms of stiffening and actuation.

The manipulator is composed of two connected identical modules, Figure 2.9. Each module possesses the original structure, with three fluidic chambers for the active omnidirectional motion combined with a central stiffening chamber, which exploits the granular jamming-based mechanism. This approach allows actuating and stiffening the modules independently.

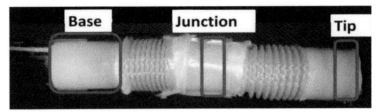

Figure 2.9   Fabricated 2-module manipulator.

2.3.1 Fabrication

The body of the module is fabricated by means of a molding process using Silicone (Silicone 0050, Ecoflex, *Smooth on Inc.*, Shore Hardness = 00-50, 100% linearized Tensile Modulus = 83 kPa). A crimped sheath is put around the module in order to contain the ballooning effect due to the chambers inflation. The detailed fabrication process of the module can be found in [13].

Each module incorporates a central channel for the granular jamming-based stiffening mechanism composed of an external latex membrane filled with 6 g of coffee; a 2 mm pipe is inserted inside and the membrane is sealed around it with Parafilm. The stiffening chamber is extended by approximately 0.5 cm on both sides with respect to the module length as shown in Figure 2.10a, top. This feature allows for keeping the stiffness variation

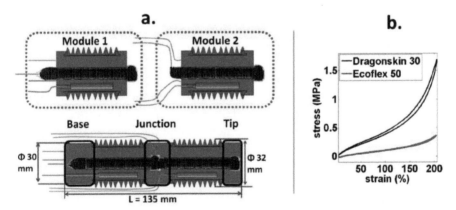

Figure 2.10   (a) Fabrication steps of the multi-module manipulator. Top, section of the two modules before connection. Bottom, two interconnected modules. In blue the pipes for the fluidic actuation; in orange the pipes for supplying vacuum to the stiffening chambers. The total length of the manipulator is given by the length of the two modules (50 mm), plus 10 mm of junction, 10 mm of the tip, and 15 mm on the base. (b) Experimental stress-strain curve of silicone rubbers according to ISO 37:2005(E).

capability even in the junction between the two modules by guaranteeing that the granular material will be present also in this section.

The pipes used both for inflating the fluidic chambers and for vacuuming the stiffening chamber are 2 mm outer diameter, 1.2 mm inner diameter polyurethane tubes (*SMC Corporation*). They are connected to the fluidic chambers after the fabrication of the modules. In order to avoid possible leakages, Sil-Poxy silicone rubber adhesive (*Smooth-on Inc.*) is used to glue them in the silicone channel used for supplying air to the fluidic chambers. In order to increase the adhesion of the pipes, their tips are scratched with sandpaper.

The inter-module connection is constructed by positioning the modules at 1 cm distance, while keeping the fluidic chambers aligned. This implies that the parts of the stiffening chamber sticking out from the module's top and bottom are in contact and slightly compressed among each other.

Two half-cylindrical shells with an inner diameter of 32 mm are then used to cap the junction and Silicone (Silicone 30, Dragonskin, Smooth on Inc., Hardness = 30, 100% linearized Tensile Modulus = 593 kPa) is poured inside. The same procedure is repeated at the top and the bottom of the manipulator, in order to fully close the structure (Figure 2.10a, bottom). The base is extended by 1.5 cm in order to simplify the clamping of the manipulator during the testing phase.

The use of a stiffer silicone material in the passive parts and particularly in the junction area guarantees that they do not affect the overall behavior of the manipulator. Dragonskin 30 silicone was chosen since it presents a stiffness 7 times higher than Ecoflex 0050 which is used for the fabrication of the modules. The mechanical properties of the two silicones were tested according to ISO 37:2005(E) and the stress-strain data are shown in Figure 2.10b. The curves of Figure 2.10b represent the average of five cycles of loading/unloading of the silicone performed with an Instron 5900 Testing System. The maximum measured variability was ±3.4 kPa for the Ecoflex 0050 and 2.8 kPa for the Dragonskin 30.

2.3.2 Workspace Evaluation

The manipulator has been characterized through experimental tests aimed at verifying its dexterity, stiffening capability, and possibility of exploiting stiffness variation during the application of forces.

2.3.2.1 Methods

The actuation of the 2-module manipulator was performed by controlling the pressure in each fluidic chamber independently. Six proportional pressure regulator valves (series K8P, E.V.P. systems) were used to modulate the air pressure inflated in each chamber from 0.0 to 0.065 MPa (inset of Figure 2.11a). A compressor (Compact 106, Fiac Air-Compressors) was used as pneumatic air source. Vacuum for stiffness modulation was obtained by a vacuum regulator (ITV0090, SMC Corporation), shown in the inset of Figure 2.11a, and a vacuum pump (LB.4, D.V.P. Vacuum Technology).

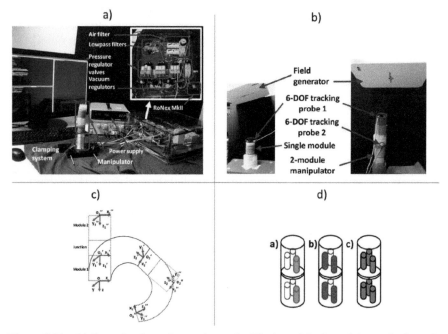

Figure 2.11　(a) Setup for the active motion and stiffening of the 2-module manipulator. In the inset the top view of the box is shown, indicating all components used for the control of the pneumatic actuation and of the vacuum levels. (b) Left, setup for the experimental measurement of the workspace of the single module. Right, setup used for the experimental measurement of specific configurations of the 2-module manipulator. The 6 DoF localization probe (Northern Digital Inc.) is highlighted in red in the pictures. (c) Scheme of the extrapolation strategy for computing the workspace of the 2-module manipulator from the workspace of the single module. The global coordinate system is o, the local coordinate system of the first module is o′ and the local coordinate system of the second module is o″. (d) Scheme of the 2-module manipulator highlighting the chambers activated for single chamber bending (a), two chamber bending (b) and elongation (c).

The vacuum pump is able to provide a maximum vacuum of 0.03 Pa absolute pressure with a flow of 3 m³/h. The vacuum generated inside the stiffening chamber was monitored with a pressure sensor (SWCN-V01-P3-2, Camozzi) and resulted in a maximum of –0.0987 MPa relative pressure. A 5 μm filter (MC104-D10, E.V.P. Systems) was used to prevent particles from entering the pump. The control of the pressure and vacuum regulators is done with low-pass filtered PWM signals generated from the digital I/O pins of the RoNex MkII (http://www.shadowrobot.com/products/ronex/).

The pressure within each chamber can be regulated by setting the period and the ON-time of the PWM signal for each pin. The RoNex MkII is programmed using the Robot Operating System (ROS). The components for the control of the manipulator are illustrated in Figure 2.11a. During the tests, the manipulator was fixed with a clamping system, as shown in Figure 2.11a.

The workspace of the 2-module manipulator was estimated through extrapolation from the single module one. The workspace of the single module was obtained by placing a 6-DOF probe (Northern Digital Inc.) on the tip of a single module and measuring the position and orientation of the tip at all the different pressure combinations in the three fluidic chambers (Figure 2.11b, left). The pressure tested in the chambers was varied from 0.0 to 0.065 MPa. Since the module motion in response to the applied pressure is not linear, the following pressures were tested, [0.00, 0.025, 0.035, 0.045, 0.050, 0.065] MPa. These pressures were experimentally found to significantly describe the motion of the module in previous works. All the different combinations of the aforementioned pressures were applied in the three chambers, thus realizing 63 combinations (i.e., 216 points). Each pressure combination was automatically set by the control system in ROS; between two pressure combinations the pressure was reset to 0.0 MPa in all the chambers. During the application of each combination of pressures, the position and orientation of the module tip were acquired with the Aurora® EM Tracking system for 1 s (i.e., 100 samples).

The workspace of the 2-module manipulator was computed from such data by considering, for each point reached from the tip of the first module, all the possible configurations of the second one. In Figure 2.11c, the procedure is shown in an exemplified scenario; the coordinate system o is the global coordinate system. The localization probe measures the position and orientation of the coordinate system o′ for each point of the workspace of the single module (with the Aurora tracking system).

The transformation matrices from the coordinate system o to o′ are computed from these experimental data. Assuming that the two modules composing the manipulator are identical, the two transformation matrices T01 (from global coordinates to first module tip) and T12 (from first module tip to second module tip) can be considered identical. The orientation and position of o″ (for every configuration) can be obtained by multiplying each T01 (one for every point of the first module workspace) for all the T01. The junction area is considered as an extra translation matrix in the local coordinate system o′. In Figure 2.11c, some explicative configurations of the system are drawn.

In configuration 1 (no actuation, rest condition), the position of the tip of the second module is obtained by a simple translation along the $z1'$ axis. When the first module is bent (configuration 2), applying the same transformation as before, the point o2″ can be obtained. Similarly, when even the second module would be bent, the same transformation that maps o into o2′ can be used on o2′ to obtain o3″.

Some relevant configurations of the manipulator, including elongation, bending with single-chamber actuation, and bending with 2-chamber actuation, were measured experimentally in order to assess the effectiveness of the behavior of the manipulator and compare it with the data obtained computationally by extending the workspace of the single module. Such measurements were performed by placing two 6-DOF probes on the manipulator, one at the tip and one at the end of the first module as shown in Figure 2.11b, right. In all cases the inflation pressures tested were [0.00, 0.025, 0.035, 0.045, 0.050, 0.065] MPa. In the case of single-chamber bending, one chamber of each module was inflated at the same time (Figure 2.11d, case a); for the 2-chamber bending, two chambers per module were pressurized by the same value and at the same time (Figure 2.11d, case b); for the elongation measurement all three chambers of the two modules were inflated by the same pressure at the same time (Figure 2.11d, case c).

2.3.2.2 Results

In Figure 2.12 the full workspace of the 2-module manipulator is shown. In Figure 2.12a, a section of the workspace is shown; the section cuts in two parts the workspace on the x-plane in order to have a clearer visualization. The initial position of the manipulator (no actuation) is schematically shown in the plot as a cylinder. The arc drawn by the points on the left side of Figure 2.12a corresponds to the single-chamber bending and is in agreement

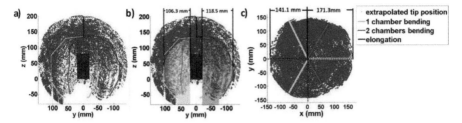

Figure 2.12 Workspace of the 2-module manipulator. (a) Section of the workspace along the x-plane, lateral view. (b) Section of the workspace with highlighted the unreachable areas. (c) Top view of the workspace.

with the results obtained with the single module in [13]; higher curvatures can be achieved and lower points in z direction can be reached (–75 mm with respect to the base of the manipulator). The right side of Figure 2.12a corresponds to the 2-chamber bending and presents a bigger radius of curvature. The experimental trajectories of the tip of the 2-module manipulator during single-chamber bending (cyan), 2-chamber bending (red) and elongation (black) match the extrapolated data. In Figure 2.12b the unreachable areas inside the workspace are highlighted; the yellow area is below the single-chamber bending and thus it extends slightly less than the red area, which is in correspondence with the 2-chamber bending. In Figure 2.12c the top view of the full workspace is reported. The planes corresponding to single- and 2-chamber bending are highlighted in cyan and red respectively. The system presented good symmetry properties (with 120° phase) in its behavior. The maximum diameter of a circle containing the whole workspace in the x-y plane (Figure 2.12c) is 312.4 mm.

In Figure 2.13 the results from the single- and 2-chamber bending are reported. Figures 2.13a and b represent the tip trajectories (position and orientation) of the manipulator during single- and 2-chamber bending respectively, together with a picture of the manipulator at the corresponding maximum reachable angle. In the plots of Figures 2.13a and b, the manipulator is reported in the non-actuated configuration as a cylinder and the trajectories, derived from the workspace extrapolation, are reported in blue, while the experimental data are in red. The two trajectories are very close for small pressures (around 0.04 MPa), but after that the error increases considerably, in particular along the z coordinate and especially for the 2-chamber bending trajectory. A very similar trend applies to the orientations (although the errors are smaller): the estimated bending angle is 236°±3.4° while the

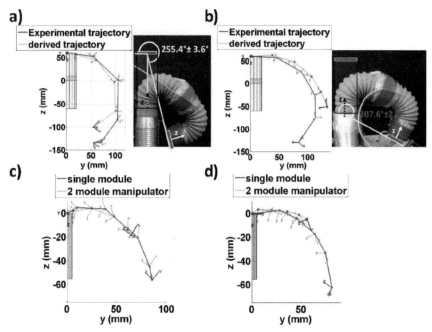

Figure 2.13 Comparison between experimentally measured trajectories end extrapolated ones. (a) Trajectories of the single-chamber bending in the 2-module manipulator and photo of the manipulator reaching the maximum bending angle with the single chamber inflation in both modules of 0.065 MPa. (b) Trajectories of the two chambers bending in the 2-module manipulator and a photo of the manipulator reaching the maximum bending angle with the 2-chamber inflation in both modules of 0.065 MPa. (c) Trajectory during single-chamber bending of a single module per se and when integrated in the 2-module manipulator. (d) Trajectory during two chamber-bending of a single module per se and when integrated in the 2-module manipulator.

measured one is $255°\pm3.6°$. In the case of the 2-chamber bending, the computed bending angle is $175°\pm1.8°$ and the experimentally measured one is $207°\pm2.3°$.

The possible reason for this difference can be found in Figures 2.13c and d where the trajectories of the single (Figure 2.13c) and 2-chamber (Figure 2.13d) bending, measured on a single manipulator (blue) and measured at the end of the first module of the 2-module manipulator (red) are reported. It is evident that the module is pushed down by the weight of the second module and this effect is higher at larger bending angles. In addition, the maximum bending angle with single-chamber bending of the single module is 118°±3.2° while in the 2-module manipulator is 132°±2.9°. Similarly, the maximum bending angle with 2-chamber bending of the single module is 87.5°±1.8° while in the 2-module manipulator it is 115°±2.2°.

The experimental data suggest that the estimated workspace will present more points at the bottom in reality, but still it is a good approximation of the manipulator reachable space.

2.3.3 Junction Characterization

2.3.3.1 Methods

The mechanical properties of the junction area between the two modules were experimentally characterized using the setup shown in Figure 2.14. The active parts of the 2-module manipulator were fully constrained with two rigid shells and a fixed displacement was imposed to the tip of the manipulator by a 6 DOF industrial robot (RV-6SL, Mitsubishi) with an F/T sensor (MINI 45, ATI, USA, resolution = 0.025 N) fixed on its end effector. In this way the overall deflection was due only to the behavior of the junction area. The test was performed at different vacuum pressures in the stiffening chamber, i.e., 0.0 MPa, –0.052 MPa, and –0.098 MPa in both the stiffening chambers of the two modules; each test was repeated three times.

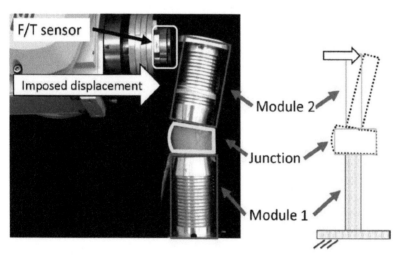

Figure 2.14 Setup for the experimental characterization of the junction are between the two modules. Left, assembled setup; in red the modules composing the manipulator are indicated, in yellow the deformed junction is highlighted. Right, scheme of the system.

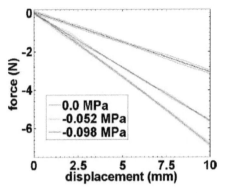

Figure 2.15   Results from the characterization of the junction between the two modules of the manipulator.

2.3.3.2 Results

In Figure 2.15, the results for the tests on the junction are reported. The forces necessary to deform the junction area reaches a maximum of 3.07±0.56 N at 10 mm displacement. Under the same conditions but applying −0.052 MPa pressure in the stiffening chambers, the force increases 5.61±0.08 N (83% increase) and reaches 6.86±0.12 N (123% increase) at −0.098 MPa. The slope of the curves (elastic constants) varies from 0.31 N/mm at atmospheric pressure to 0.54 N/mm at −0.052 MPa and 0.69 N/mm at −0.098 MPa.

2.3.4 Stiffness Characterization

2.3.4.1 Methods

The stiffening capabilities of the single module in different configurations in terms of bending and elongation have been extensively characterized in [13]. Here, the stiffening capabilities of the manipulator as a whole are presented. Tests were performed imposing different displacements at the tip of the manipulator by using a 6 DOF industrial robot (RV-6SL, Mitsubishi) with an F/T sensor (MINI 45, ATI, USA, resolution = 0.025 N ATI Mini45) fixed upon its end effector. In that way, it has been possible to impose the right orientation of the load cell respect to the module tip position. The same test was performed when the stiffening mechanism was not activated (0.1 MPa) and when −0.1 MPa vacuum was induced in the granular jamming-based stiffening mechanism; each test was repeated five times. The stiffness variation was characterized in both compression and tensile tests. Compression tests were performed compressing the manipulator along the z direction

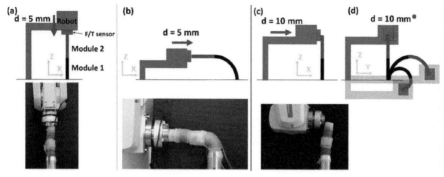

Figure 2.16   Manipulator configuration tested for evaluating the stiffening capabilities. The red square represents the F/T sensor, the blue structure represents the robot end effector; the base module of the manipulator is the black one and the second module is the dark gray one. (a) Axial test, (b) axial test with the first module 90° bent, (c) side view and (d) front view of the lateral test at different bending angles of the manipulator. For each configuration, the photo of the real setup is reported below.

(Figure 2.16a) of 5 mm. The same compression test was performed even when the first module is 90° bent (Figure 2.16b). Tensile tests were performed imposing a lateral displacement to the manipulator. Such tests were carried out both when the manipulator was in the fully straight configuration (Figure 2.16c) and at different bending angles (Figure 2.16d). The tests at different bending angles were performed inflating one chamber on each module with the same pressure; the tested pressures were [0.25, 0.35, 0.45, 0.55, 0.65] bar (Figure 16d).

2.3.4.2 Results

The results from the stiffness tests of Figure 2.16 are presented in Figure 2.17. The plots report the force measured from the F/T sensor with respect to the imposed displacement of the manipulator. Three different vacuum levels were applied to the stiffening chamber in order to verify the possibility of tuning the stiffness level. In Figure 2.17 (left) the results correspond to the configuration of Figure 2.16d. The force necessary to deflect the manipulator visibly changes according to the stiffening level. As an indication of the stiffness, the elastic constant was computed as the slope of the linear tract of the curves for the first 3 mm of displacement. The elastic constant varies from 0.11 N/mm when no stiffening is activated to 0.20 N/mm at –0.05 MPa vacuum pressure and up to 0.31 N/mm at –0.1 MPa.

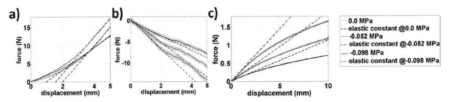

Figure 2.17   Results from the stiffness tests. (a) Axial tests. (b) Axial tests with the first module bent 90° (0.045 MPa inflation on one chamber). (c) Lateral test with no chamber inflation.

In Figure 2.17 (center) the results from the tests in the axial direction are reported (Figure 2.16a). In this case, for the first 2.5 mm displacement the effect of the stiffness variation is not evident. This is probably due to the change in volume of the stiffening chamber that tends to pack the granules together, thus leaving the tip with less granules and with a stiffness similar to the silicone one. On the other hand, it is possible to appreciate the stiffness variation when the displacement increases over 3 mm. In this case, the elastic constant was computed as the slope of the curves in the last part of the plot. The elastic constant varies from 2.18 N/mm when no stiffening is activated, up to 3.15 N/mm at –0.05 MPa vacuum pressure and 5.1 N/mm at –0.1 MPa.

Figure 2.17 (left) shows the results from the axial test when the first module is bent of 90° (Figure 2.16b). In this case the elastic constant varies from 1.99 N/mm when no stiffening is activated, to 2.6 N/mm at –0.05 MPa vacuum pressure, and 2.96 N/mm at –0.1 MPa. It is interesting to observe that the manipulator is able to withstand relatively high forces, also in the bent configuration; in particular, it withstood up to 17 N at –0.1 MPa vacuum pressure that is relevant for surgical tasks. In addition, these last curves presented a higher variability and a peak at around 3 mm. In fact, above a certain force, it starts to separate the jammed granules of the stiffening chamber and thus the performance of the stiffening mechanism decreases.

The results from all the stiffening tests are summarized in Table 2.1. The change in the elastic constant in the experiments was computed in the configurations shown in Figure 2.16. In the last column, the percentage of change in the stiffness is computed. It is important to observe that the stiffness variation is maintained also during the bending of the manipulator. As evident from Table 2.1, the elastic constant decreases due to the bending of the structure; however, the stiffening mechanism guarantees in all the configurations tested a considerable increase in the stiffness of the manipulator.

Table 2.1 Summary of the results from the stiffening tests on the 2-module manipulator

Test Typology	Chamber Inflation Pressure (bar)	Elastic Constant (N/mm) @ 0.1 MPa Pressure (Stiffening Chamber not Active)	Elastic Constant (N/mm) @ –0.1 MPa Pressure (Stiffening Chamber Active)	Elastic Constant Increase (%)
Lateral	0	0.11	0.31	66.6
	0.25	0.08	0.14	75
	0.35	0.05	0.15	200
	0.45	0.06	0.17	183,3
	0.55	0.06	0.17	183,3
	0.65	0.07	0.14	100
Axial	0	2.18	5.10	133.9
	0.45 @module1	1.99	2.96	48.7

2.3.5 Combined Force and Stiffening Experiments

2.3.5.1 Methods

Two different types of tests were carried out to evaluate the forces exerted by the manipulator exploiting the selective stiffening capabilities of its segments. The first test consisted of positioning the manipulator in the same configuration as for the stiffening tests (Figure 2.16c). Three different vacuum levels ([0, –0.05, –0.1] MPa) were imposed in the first module (in black in Figure 2.8c) and the chamber of the second module that causes a bending on the x-z plane of Figure 2.16c was inflated at [0.25, 0.35, 0.45, 0.55, 0.65] bar. Forces were measured using an F/T sensor (ATI Nano17). The same procedure was performed while stiffening the second module and actuating the first one.

The second type of tests were aimed at exploiting the stiffening capabilities together with the possibility to generate forces in a more surgery-like scenario. Although the previously described tests provided a good overview of the 2-module manipulator, they still lack a thorough demonstration of the real capabilities of such a structure, in comparison with traditional rigid-link surgical manipulators.

For that reason, scenarios like that proposed in the schematic view of Figure 2.18 have been taken as guidelines to build a more reliable, credible test setup for the 2-module manipulator. To reproduce the compliance, in terms of weight and shape of organs or anatomical parts that the manipulator may encounter during surgical laparoscopic procedures, water-filled balloons have been employed and they have been placed around the manipulator to test its interaction with such objects.

Among the variety of possible tasks, a few key movements were chosen to demonstrate the manipulation and stiffening capabilities. These are the wrapping and retraction of a water-filled balloon (500 g, Figure 2.19a), hung

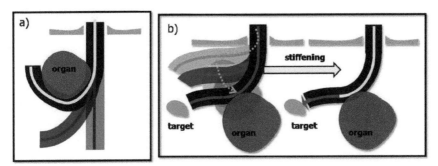

Figure 2.18 Schematic examples of surgical tasks performed by a tentacle-like structure. Left: organ retraction, showing the manipulator grabbing and lifting up of the organ. Right: fitting in tiny spaces, shifting down of an organ with the base portion, and reaching the surgical target with the distal module. The yellow line indicates the stiffening of the manipulator portion.

up to a load cell which revealed when the whole weight of the balloon was supported by the manipulator. Another task is shown in Figure 2.19 (center) where the manipulator navigates among compliant objects (water-filled balloons), embraces one of them (270 g), and moves it aside. The last

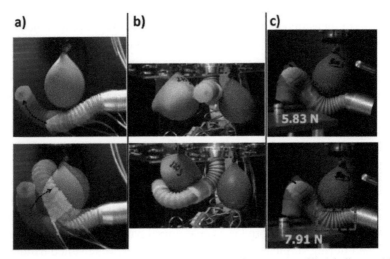

Figure 2.19 (a) the 2-module manipulator passes below a water filled balloon with the first module and exploits the second module to grasp and move the baloon around; (b) the manipulator fits between two water-filled balloons, lifting and shifting one of them to free access to the other one; (c) the manipulator is able to keep the weight of a 500 g balloon with the first module and apply a variable force on the F/T sensor.

task, presented in Figure 2.19 (right) demonstrated the manipulator supporting a weight of 500 g with the first module and applying a force on an F/T sensor. The same test is performed without stiffening activation and when the first module is fully stiffened. In this last experiment, two F/T sensors were used. One F/T sensor is connected to the water-filled balloon (5 N), while the other is positioned in the proximity of the distal end of the manipulator. In this test, two pieces of information can be extracted. The first F/T sensor allows verifying that the water filled balloon is completely supported, while the second F/T sensor measures the amount of force generated on the target.

2.3.5.2 Results

In Figure 2.20, the results from the combination of stiffening and actuation are presented. On the left the case when the stiffness of the base module is changed and the top module applies forces to the F/T sensor is shown. In this case the maximum force exerted when no stiffening is activated tends to saturate at approximately 1 N. On the other hand, when the base module is stiffened, the force is transmitted more effectively since it creates a more stable support for the second module when it applies force to the F/T sensor. The maximum force when the base module is fully stiffened reaches 2.2 N. This feature is important in order to apply force in a controlled way to tissues and biological structures. In the absence of stiffening capability, if the force necessary to shift a weight is too high, the structure may not succeed and may deform in an uncontrolled way in other directions due to its highly compliant structure.

The same test was performed by actuating the first module and changing the stiffness in the first one. In this case, the effect of the stiffness variation is not as effective as in the previous case. This is probably due to the fact

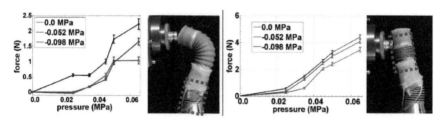

Figure 2.20   Results from the testing on the combination of actuation and stiffening. (Left) Stiffening of the base module, highlighted with the dashed square, and actuation of the first one. (Right) Stiffening of the second module (highlighted with the dashed square), and stiffening of the second one.

that the stiffening system is integrated in the central channel in order to keep the external compliance of the robot and thus its effect is mediated from the soft material in between. However, the maximum transmitted force increases from 3.45 to 4.47 N when the stiffening system is activated.

In Figure 2.19 some of the 2-module manipulators interacting with water-filled balloons are reported. In Figure 2.19c, in particular, the manipulator was able to keep the weight of a 500 g balloon with the first module and apply a force on the F/T sensor. The same test was performed when the first module was not stiffened and when it was stiffened. In the first case, the maximum recorded force was 5.83 N, in the second 7.91 N, thus validating the results obtained in Figure 2.19, left.

References

[1] Chang, B., Chew, A., Naghshineh, N., and Menon, C. (2012). A spatial bending fluidic actuator: fabrication and quasi-static characteristics. *Smart Mater. Struct.* 21:045008.

[2] Webster, R. J. III., and Jones, B. A. (2010). Design and kinematic modeling of constant curvature continuum robots: a review. *Int. J. Rob. Res.* 29, 1661–1683.

[3] Suzumori, K., Iikura, S., and Tanaka, H. (1991). "Flexible microactuator for miniature robots," in *Proceedings of the Micro Electro Mechanical Systems, MEMS '91, An Investigation of Micro Structures, Sensors, Actuators, Machines and Robots* (Nara: IEEE), 204–209.

[4] Greef, A. D., Lambert, P., and Delchambre, A., (2009). Towards flexible medical instruments: review of flexible fluidic actuators. *Precis. Eng.* 33, 311–321.

[5] Suzumori, K., Maeda, T., Watanabe, H., and Hisada, T. (1997). "Fiberless flexible microactuator designed by finite-element method," in *Proceedings of the IEEE/ASME Transactions on Mechatronics* (Roma: IEEE), 281–296.

[6] Suzumori, K., Endo, S., Kanda, T., Kato, N., and Suzuki, H. (2007). "A bending pneumatic rubber actuator realizing soft-bodied manta swimming robot," in *Proceedings of the IEEE International Conference on Robotics Automation 2007* (Roma: IEEE), 4975–4980.

[7] Cheng, N. G., Lobovsky, M. B., Keating, S. J., Setapen, A. M., Gero, K. I., Hosoi, A. E., et al. (2012). "Design and analysis of a robust, low-cost, highly articulated manipulator enabled by jamming of granular

media," in *Proceedings of the IEEE International Conference Robotics and Automation 2012 (ICRA)* (Saint Paul, MN: IEEE), 4328–4333.

[8] Brown, E., Rodenberg, N., Amend, J., Mozeika, A., Steltz, E., Zakin, M. R., et al. (2010). Universal robotic gripper based on the jamming of granular material. *Proc. Natl. Acad. Sci. U.S.A.* 107, 18 809–18 814.

[9] Steltz, E., Mozeika, A., Rembisz, J., Corson, N., and Jaeger, H. M. (2010). "Jamming as an enabling technology for soft robotics," in *Proceedings of the SPIE Conference on Electroactive Polymer Actuators and Devices 2010*, San Diego, CA.

[10] Loeve, A. J., van de Ven, O. S., Vogel, J. G., Breedveld, P., and Dankelman, J. (2010). Vacuum packed particles as flexible endoscope guides with controllable rigidity. *Granul. Matter* 12, 543–554.

[11] Jiang, A., Ataollahi, A., Althoefer, K., Dasgupta, P., and Nanayakkara, T. (2012). "A variable stiffness joint by granular jamming," in *Proceedings of the ASME 2012 International Design Engineering Technical Conferences and Computers and Information in Engineering Conference IDETC/CIE 2012*, Chicago, IL.

[12] Kaufhold, T., Bohm, V., and Zimmermann, K. (2012). "Design of a miniaturized locomotion system with variable mechanical compliance based on amoeboid movement," in *Proceedings of the 4th Biomedical Robotics and Biomechatronics (BioRob) 2012 IEEE RAS and EMBS International Conference on IEEE RAS* (Rome: IEEE), 1060–1065.

[13] Cianchetti, M., Ranzani, T., Gerboni, G., De Falco, I., Laschi, C., and Menciassi, A. (2013). "STIFF-FLOP surgical manipulator: mechanical design and experimental characterization of the single module," in *Proceedings of the IEEE/RSJ International Conference on Intelligent Robots and Systems*, Tokyo, 3576–3581.

3

Soft Manipulator Actuation Module – with Reinforced Chambers

Jan Fras[1], Mateusz Macias[1], Jan Czarnowski[1], Margherita Brancadoro[2], Arianna Menciassi[2] and Jakub Głowka[1]

[1]Industrial Research Institute for Automation and Measurements PIAP, Warsaw, Poland
[2]The BioRobotics Institute, Scuola Superiore Sant'Anna, Pontedera (PI), Italy

Abstract

During the integration and data-fusion algorithms tests it turned out that the initial manipulator prototype, described in Chapter 2, had a number of drawbacks emerging from its structure and resulting in high actuation non-linearity and low sensing capabilities. In this chapter we show how the internal distortions in the initial design influence the manipulator behaviour and the sensory readings, and then we propose new designs that solve those issues. A braided actuation chamber instead of the overall module braiding concept is presented together with the manufacturing technology and experimental characterization of the actuation module. The idea of using multiple actuation chambers and the division of the central stiffening chamber into distributed stiffening chambers is also presented.

3.1 Introduction

The braided sleeve used for the initial design of the STIFF-FLOP manipulator rendered itself to be a good solution (Chapter 2). The actuation chamber's radial expansion were not being observed and the modules were still able to bend, elongate, and squeeze. The problem seemed to be solved; however, the sensors integration process has shown that the external inflation of the chambers during actuation is a bit more complex and the braided reinforcement limits only the symptoms of the problem. Moreover, since such a solution limits the external expansion, it intensifies the deformation of the actuation chambers in an inward direction. In addition, the interaction between the silicone body of the manipulator and the external reinforcement structure introduces some new undesired and unexpected effects. Effects that have been noticed are: non-linear actuation, distortion of the internal sensors readings (mainly the curvature and length sensors as they are based on the structural deformation of the manipulators' body), complex control, and modeling.

In the following sections we analyze those effects and their relation to the initial braiding solution. The analysis is derived from several observations of the manipulators' behaviour during actuation tests. For that a dedicated module has been used. One of its ends was cut and sealed with a transparent tile that allowed for the observation of its interior.

3.1.1 Change of the Chamber Cross Section Area

As the actuation chamber is simply an empty cylindrical channel in the silicone material, application of pressure into it causes it to expand radially. That effect influences the cross section area of the activated actuation chamber and the area of the other chambers' cross sections as well. As the force resulting from pressure acting inside the chamber, depends not only on the pressure but also on the cross section of the chamber the mentioned increase of the chamber dimensions causes the force to change its value in a non-linear way as a function of pressure.

3.1.2 Chamber Cross Section Center Displacement

In the initial design the cross section of the actuation chamber was semicircular. Its shape is not retained during the actuation due to the material deformation constrained by the braiding and the overall module cross section geometry. That causes the chamber cross section geometry and its location to change when the pressure changes. Consequently the position of the actuation cross section moves in relation to the manipulator central axis depending on the pressure inside the chamber. Actuation of two chambers simultaneously results in a lower bending angle than the single-chamber actuation with the same pressure. The cause of this situation is presented in Figure 3.1. The explanation is that the bending deformation of the manipulator results from

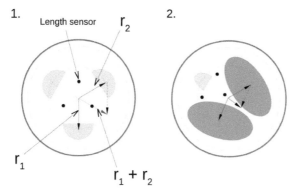

Figure 3.1 Deformation of the module cross section during actuation. The two-chamber actuation case results in their expansion and their axis displacement that results in lowering the pressure-dependent bending moment in the cross section.

bending moment generated by actuation chambers. This in turn is caused by the forces resulting from internal actuation pressures. As the pressure value p is constant at every point of the chamber volume, the value of the internal force acting on a certain chamber cross section center can be easily calculated as $F_i = A_i p$, where A_i is the area of the cross section [1]. Denoting vectors from the center of the cross section to centers of each chamber as $\vec{r_1}$ and $\vec{r_2}$, the resulting bending moment M in the considered cross section, the module can be expressed as Equation (3.1).

$$M_i = (\vec{r_1} + \vec{r_2})F_i \qquad (3.1)$$

In Figure 3.1, the situation with two simultaneously actuated chambers is presented. The shape of the chamber cross section is deformed and its geometrical center moves. Due to the passive module geometry, the chamber cross section shifts towards the module center and as a result the length of the net vector $\vec{r} = \vec{r_1} + \vec{r_2}$ is decreased. That in turn leads to a reduction in the value of the net bending moment deforming the module Equation (3.1).

3.1.3 Friction between the Silicone Body and Braided Sleeve

As mentioned in the chapter introduction, the braided sleeve, beside constraining the ballooning effect, also influences other manipulator properties. The final shape of the manipulator depends on the previously applied pressures and in particular on the chamber actuation order. The friction force caused by the first chamber actuation limits the second chamber expansion

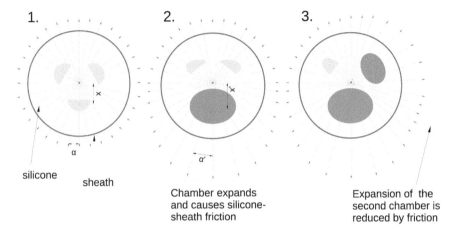

Figure 3.2   Module cross section during a two-module actuation sequence. (1) presents the cross section with no pressures applied to the chambers. No friction between the sheath and silicone occurs. In (2), pressure is applied to one of the chambers. The chamber expansion causes the external sheath to make contact with the module body. The second chamber is constrained by the inflation and friction with the sheath (3).

and reduces its elongation and thus its impact on the manipulator. This effect is clearly observed in case of two chambers subsequent actuation Figure 3.2. Applying the pressure to a chamber makes it expand and increase its cross section. Due to that, the module surface is pressed to the reinforcement from the inside. When the second chamber is actuated, its expansion is significantly smaller than the first one since the space for the expansion is already occupied by the initially pressurized chamber that was expanding without such constraints. Pushing it away requires moving the silicone pressed to the braiding, which requires additional force due to the friction. Consequently the second chamber grows less than the first one and due to this, the pressure-related elongation is also smaller. As an effect, the module is not capable of efficient bending in the directions requiring multiple chambers to be actuated, which greatly impacts the shape of an achievable workspace when using constant pressure.

3.1.4 Sensor Interaction

One of the sensors developed in the STIFF-FLOP project used for manipulator configuration detection is a curvature sensor composed of three parallel-length sensors [2], that are made of a light fiber slighting in a dedicated

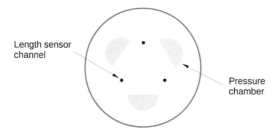

Figure 3.3 The length sensor channel placement presented in a module cross section. The channels are placed on a circle coaxial with the module axis of symmetry.

hollow. Each individual fiber is attached at the modules' tip so that any elongation of the module causes the fibers to be pulled inside its body. That in turn makes the end of the fiber change its distance form a light intensity detector and finally allows the lenght of the particular part of the module to be calculated. The sensors are embedded into the module with axial symmetry Figure 3.3. and are assumed to be parallel to the module central axis. Any deviation from that is a source of error. The internal inflation of the chambers disrupts this assumption, as the readings from the sensors are highly dependent on the pressure and not only on the shape of the manipulator. The curvature of the module is assessed using the individual lengths of the sensors, and so that is a significant issue and makes sensor fusion a complex task.

Another issue is that the braiding structure surrounding the manipulators' body is undesired due to its roughness, which may cause harmful interaction with the patients' tissue and an array of tactile sensors that was considered to be integrated with the manipulator's surface. Integration of such sensors rendered itself to be much easier done in case of the homogenous silicone body itself than in the braided reinforcement. More information on this type of sensors can be found in [3].

3.2 Proposed Improvements

To reduce the issues described in the introduction, radial expansion of each individual chamber has to be constrained. Any changes to the cross section geometry are undesired too, as they may result in a less linear actuation process. The idea we propose is to make the chamber geometry more stable by employing braiding around each chamber instead of constraining the external expansion of all chambers with one external sheath.

3.2.1 Possible Solutions

The external braided sleeve embedded to solve the undesired module expansion demonstrated to provide effective radial expansion constrains allowing the longitudinal chambers expansion at the same time, but due to a number of undesired effects another solution has to be proposed. Similar examples of constraint can be found in the work by Whitesides et al. who, facing the same ballooning effect [4], developed a manufacturing process which allows the internal patterning of the chambers (PneuNets), reducing the lateral space available to produce outward expansion [5]. A different and simpler approach has been proposed by Brock et al. Their solution is a squared-section chamber reinforced with inelastic yarns placed all around [6]. Such an approach is simpler and faster to manufacture, and provide a similar performance to the first one.

3.2.2 Design

The first solution used semi-cylindrical chambers. This approach would ensure that the chamber circumference is constant during the actuation, and so limit the radial expansion. In general, the cross section geometry would, however, still not be constant as the chamber behaviour during actuation will aim to maximize its cross section area, and will finally reach a circular cross section shape. This effect has been discussed in more detail in [7]. Therefore in the proposed solution, the semicylindrical chambers have been substituted with cylindrical ones [8]. Such a change would also simplify the manufacturing process. The concept is presented in Figure 3.4.

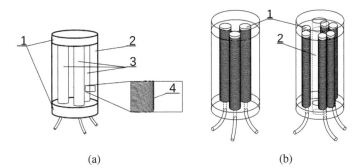

(a)   (b)

Figure 3.4   Improved design concept. (a) the module overview, 1 - top and bottom of the module made of stiff silicone, 2 - module body made of soft silicone, 3 - actuation chambers, 4 - helical thread reinforcement, (b) 1 - actuation chambers, 2 - empty central channel.

The final version of the manipulator was built using pairs of chambers working together and an empty channel for passing the pressure pipes to modules stacked one on top of another Figure 3.4(b). The requirement of empty space inside the module forced the chambers to have a smaller actuation volume and experiments have shown that such a change requires higher pressures and significantly reduced final motion capabilities of the manipulator. In order to preserve the desired actuation area, the singular actuation chambers were substituted with multiple chambers connected inside the module [9]. For the same reason the stiffening chamber has been split to three smaller channels located between the actuation chambers. Such an operation allowed to create the central channels, but also improved the stiffening performance as its area moment of inertia increased. An assembled module is presented in Figure 3.5(c) and a modified module with a central hollow channel, doubled actuation chambers, and three individual stiffening chambers is presented in Figure 3.6.

(a)   (b)   (c)

Figure 3.5 Initial manufacturing approach. (a) chambers manufacturing - winding a thread on prefabricated silicone cylinder, (b) body moulding, (c) imperfections caused by the stresses introduced during winding process.

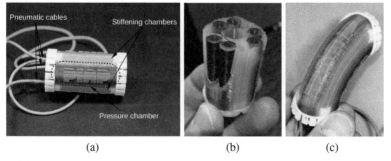

(a)   (b)   (c)

Figure 3.6 The STIFF-FLOP module. (a) overview, (b) actuation chambers disclosed, reinforcing thread visible, (c) actuated module.

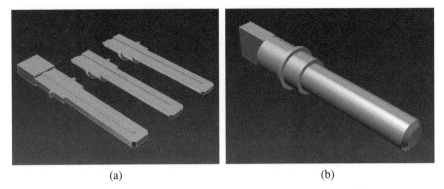

Figure 3.7   Mold core. (a) parts of the core, (b) core assembly.

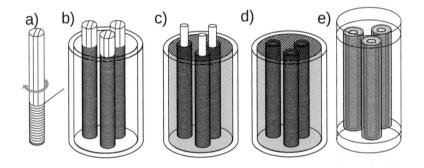

Figure 3.8   Module manufacturing steps. (a) Thread application on a rod, (b) body part molding, (c) replacement of the initial rod with a smaller-diameter one, (d) internal chamber layer created, (e) sealing of the manipulators' tip and bottom.

3.3 Manufacturing

The improved manipulator is manufactured in several molding steps. For that purpose, a set of 3D printed moulds is used. The process has been evolving during the project but the idea behind it remained the same. The main issue is to create sealed cylindrical chambers with the reinforcement incorporated into their walls and to embed them into the manipulator body. Originally the thread was applied onto prefabricated silicone cylinders (Figure 3.5(a)) and such cylinders were then sealed with the silicone material forming the module (Figure 3.5(b)).

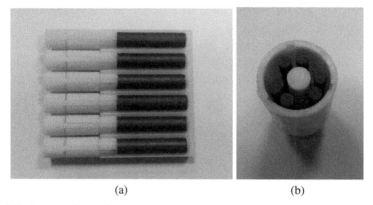

(a) (b)

Figure 3.9 Improved manufacturing approach. (a) molding cores wrapped with reinforcing thread, (b) wrapped cores in the main body mold.

Such a method was very imprecise due to the fact that the soft structure that the thread was applied on and the tension required for the proper thread placement introduced a lot of stresses into the silicone material. As a result the material was pressed in the direction of the thread application and caused many irregularities in the reinforcement structure. This effect can be observed in Figure 3.5(c) as a small difference in the pitch of the reinforcement. The material was pressed out between the reinforcement cycles. Another issue is that the friction of the chamber pressed by the applied thread made it impossible to remove the core from inside without damaging the chamber structure and forced us to solve that issue by designing a three-part core that is removed part by part Figure 3.7. The central part is removed as the first one, while the other parts protect the silicone layer. After the internal part is removed, the remaining parts can be removed without any resistance.

Due to the above issues, the technology has been improved. The manufacturing steps have been reordered as presented in Figure 3.8. The initial step is to create the reinforcement on a rigid rod (Figure 3.9). Since the rod is rigid, the thread does not introduce any significant stress into the structure of the chamber. Moreover, the wrapping process is very simple and can be easily automated, as every cycle of the reinforcement stacks onto the previous one. To simplify the core removal that is done in the next steps, the core construction has ben inherited from the previous manufacturing approach and consists of three parts. When the core is wrapped, it is inserted into the main mold and the manipulator's body is created. Once the silicone is cured, the cores are removed—part by part, the internal part at first, and then the sides, so that the thread remains attached to the body. After that, the hollow chambers with the thread attached to their walls are filled with uncured silicone and a set of smaller cores is inserted inside the mold. This is for creating a layer of silicone that corresponds with the prefabricated silicone cylinders from the initial approach. After curing, the structure of the main part of the module is ready. For the last step the top and the bottom of the module is closed with a stiff silicone.

3.4 Tests

The new version of the module has been tested in order to compare its performance with the previous design. The assumed improvements has been observed. In particular the ballooning effect has been successfully limited, while the internal structure deformation was not observed.

3.4.1 Pneumatic Actuation

Pneumatic actuation was the primary actuation method in the STIFF-FLOP project. Thus, both designs have been tested in terms of the bending angle as a function of pressure. Single- and 2-chamber actuation scenarios have been examined. The test setup is presented in Figure 3.10. The plots from Figures 3.11(a) and 3.11(b) present the results for the old and new design, respectively. The test shows that the new design works with a wider range of actuation pressures for a similar range of bending angles. This effect is caused by the efficient limitation of the ballooning effect. In the initial design the chambers were able to extend radially, which resulted in the cross sectional

Figure 3.10   Test configuration for measuring module bending by applying pressure into module chambers.

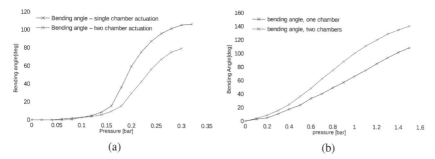

Figure 3.11 Bending angle achieved for certain values of pressure applied for: (a) the old design, (b) the new design.

actuation chamber area growth. Such a growth was causing the actuation force to be higher for the same pressure value. That resulted in greater sensitivity with increasing pressure, while ideally, with non-expanding chambers, the relation between the achieved bending angle and pressure applied to the chamber should be linear [10]. The observed improvement in terms of linearity confirms the successful limitation of the internal ballooning effect.

Another observed difference is that for the previous design, the resulting bending angle values are lower in case of 2-chamber actuation than in the case with single-chamber actuation. The reason is probably the effect described in Section 3.1.2. The new design does not display such behaviour—for the same pressures applied, the bending-angle values for 2-chamber actuation are higher than one-chamber actuation which meets our expectations.

Figure 3.12 Test configuration for measuring module bending by injection of a certain amount of incompressible liquid.

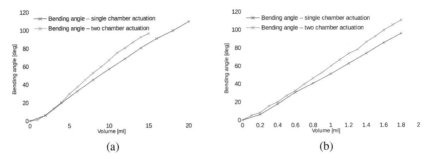

(a)   (b)

Figure 3.13   Bending angle achieved for certain values of liquid volume injected into chambers of: (a) the old design, (b) the new design

3.4.2 Hydraulic Actuation

Both designs have also been tested in the hydraulic actuation scenario. The main difference in such case is that it is not the pressure but the volume of the actuation chamber that is controlled. Such a test quantitatively shows the ballooning effect observed in the previous design. The test configuration used for this experiment is presented in Figure 3.12. The bending angle as a function of volume injected into the actuation chamber for the previous and the new design are presented in Figures 3.13(a) and 3.13(b), respectively. It can be observed that drastically lower volumes of liquid are required by the new module to achieve a certain bending-angle. The relative increase of volume to achieve certain bending angle values is approximately ten times greater for the previous module (2500% compared to 250% for the bending angle of 110 degrees).

3.4.3 External Force

In order to assess the reinforcement impact on the passive behaviour of the manipulator, the bending angle and the elongation caused by the application of external force has been measured for two similar modules: with and without braiding around the actuation chambers. The test configurations are presented in Figures 3.14(a) and 3.14(b) and the resulting plots in Figures 3.15(a) and 3.15(b). The experiment shows that there are no significant differences between those two modules, suggesting that the influence of the reinforcement on the manipulator behaviour under external forces can be neglected.

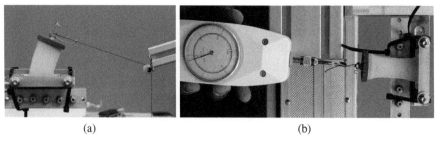

(a)   (b)

Figure 3.14 (a,b) Test setup for measuring the individual-chamber braiding influence on module bending elongation by external force, respectively.

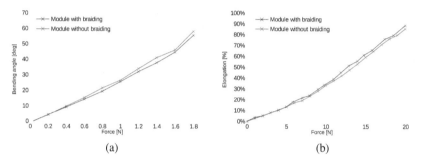

(a)   (b)

Figure 3.15 (a,b) Bending angle and elongation respectively, achieved by the modules with and without individual-chamber reinforcement when applying external force.

3.5 Stiffening Mechanism

The design of the final arm has been developed starting from the new approach of manufacturing described in this chapter. This method of module fabrication allows obtaining chambers that can only elongate in response to the pressure, as they have a circular cross section and they are enveloped by an inextensible thread wrapped around their walls. This thread is completely embedded into the matrix of the silicone. Such design has been used to build the modules composing the STIFF-FLOP manipulator; in particular two versions of chamber arrangements (i.e., basic and optimized) have been developed and integrated with the control system. The employed stiffening mechanism is the granular jamming-based mechanism (described previously); in order to leave a free lumen in the module for the passage of the pipes from and to other modules, the granular jamming chambers are distributed around the centre, similar to the actuation chambers.

3.5.1 Basic Module Design

The basic version is composed of three actuation chambers (with integrated thread) and three stiffening chambers, organized in an alternating fashion along the cylindrical STIFF-FLOP module wall parallel to the longitudinal axis of the module. The section view in Figure 3.16 (left) shows the equal distribution of all the chambers around the central axis of the module. The stiffening chambers host the particles for the granular jamming mechanism enclosed in a custom-made latex membrane. Overall, each module has three fluidic pipes (positive air pressures) that feed air into the actuation chambers (inflation/deflation) for the purpose of actuation and another three pipes to allow controlling the vacuum levels in the stiffening chambers. According to this design a three-module manipulator has been built and integrated with force sensors (refer to Part II of this book). This manipulator is composed of two modules with an integrated stiffening mechanism (I. and II.) and one without (III.), as shown in Figure 3.16 (right).

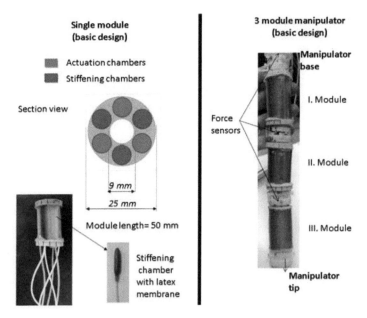

Figure 3.16   The basic module design: Left: section view of the chambers distribution and integration of the granular jamming chamber with latex membrane. Right: assembled 3-module manipulator. The total number of pipes from the manipulator is equal to 15 (6 for I. and II. module and 3 for the III. module).

3.5.2 Optimised Module Design

In order to have more efficient actuation stiffening and a minimal total number of pipes per module, an optimized version of the single module design has been developed. To maximize the effective area of the input pressure, still keeping the same size of the inner lumen and circular chamber cross section, two cylindrical actuation chambers per DoF are used. These two chambers are internally connected: so even if this module has six actuation chambers, the actuation pipes required for the pressurization of the module are still equal to three (since two chambers are always actuated at the same time). The cross section area of the stiffening chambers has been changed from a circular shape into the shape depicted in Figure 3.17 (section view), which allows occupying the maximum volume left in the module for the allocation of the stiffening mechanism. A stiffening chamber experiences a collapse of the lateral surface when a negative pressure (vacuum) is applied to the chamber, therefore a shape like the one in the cross section of Figure 3.17 (left) will assist the effect of the vacuum on the granular media by collapsing

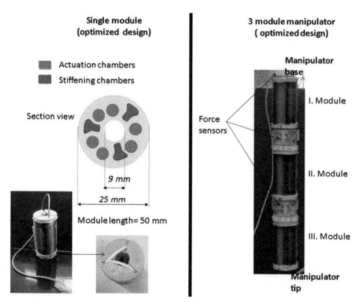

Figure 3.17 The optimised module design: Left: section view of the chambers distribution and single module with integrated granular jamming chambers without membrane and connected together. Right: assembled 3-module manipulator. The total number of pipes from the manipulator is equal to 11 (4 for I. and II. module and 3 for the III. module).

the chamber walls on the granules. In this new optimized design, the granules for jamming are not enclosed in a latex membrane anymore, but are directly in contact with the chamber's wall in the silicone matrix. Furthermore, the three stiffening chambers have been connected all together (see Figure 3.17 (left) module section view) because the vacuum is always applied simultaneously to all the stiffening chambers. The result is a module which has three actuation pipes for motion/bending and just one pipe for controlling the stiffness. With such design, a 3-module manipulator has been integrated with force sensors and interfaced with the control system for tests.

3.6 Conclusions

The new design proved its ability to successfully limit all the issues observed in previous module designs. This goal has been achieved by redesigning the actuation chamber and moving the reinforcement from outside the module to around each individual chamber. This allows for easier data fusion and modeling, and influences positively other areas of the STIFF-FLOP projects. Such a solution is not application-specific and similar actuators have been embedded in other soft robotics systems for manipulation, grasping, prosthetics and locomotion [7, 10, 11].

Acknowledgement

The work described in this chapter is supported by the STIFF-FLOP project grant from the European Commission's Seventh Framework Programme under grant agreement 287728. This project is also partly supported from funds for science in the years 2012–2015 allocated to an international project cofinanced by Ministry of Science and Higher Education of Poland.

References

[1] Fraś, J., Czarnowski, J., Maciaś, M., and Główka, J. (2014). *Static Modeling of Multisection Soft Continuum Manipulator for Stiff-flop Project.* Berlin: Springer.

[2] Searle, T. C., Althoefer, K., Seneviratne, L., and Liu, H. (2013). "An optical curvature sensor for flexible manipulators," in *Proceedings of the Robotics and Automation (ICRA)*, Hong Kong, 6–10.

[3] Sareh, S., Jiang, A., Faragasso, A., Noh, Y., Nanayakkara, T., Dasgupta, P., et al. (2014). "Bio-inspired tactile sensor sleeve for surgical soft manipulators," in *Proceedings of the 2014 IEEE International Conference on Robotics and Automation (ICRA)*, Hong Kong, 1454–1459.

[4] Martinez, R. V., Branch, J. L., Fish, C. R., Jin, L., Shepherd, R. F., Nunes, R. M. D., et al. (2013). Robotic tentacles with three-dimensional mobility based on flexible elastomers advanced materials. *WILEY-VCH Verlag* 25, 205–212.

[5] Shepherd, R. F., Ilievski, F., Choi, W., Morin, S. A., Stokes, A. A., Mazzeo, A. D., et al. (2011). Multigait soft robot. *Proc. Natl. Acad. Sci. U.S.A.* 108, 20400–20403.

[6] Deimel, R., and Brock, O. A. (2014). "Novel type of compliant, underactuated robotic hand for dexterous grasping," in *Proceedings of the Robotics: Science and Systems*, Berkeley, CA.

[7] Fras, J., Noh, Y., Wurdemann, H. A., and Althoefer, K. (2017). *Soft Fluidic Rotary Actuator with Improved Actuation Properties*. Rome: IEEE.

[8] Fraś, J., Czarnowski, J., Główka, J., and Maciaś, M. (2017a). *Soft Manipulator*. PL Patent PL226535.

[9] Fraś, J., Czarnowski, J., Główka, J., and Maciaś, M. (2017b). *Soft Manipulator Module*. PL Patent PL225988.

[10] Fraś, J., Czarnowski, J., Maciaś, M., Główka, J., Cianchetti, M., and Menciassi, A. (2015). "New stiff-flop module construction idea for improved actuation and sensing," in *Proceedings of the International Conference on Robotics and Automation* (Rome: IEEE), 2901–2906.

[11] Fraś, J., Maciaś, M., Czubaczyński, F., Sałek, P., and Główka, J. (2016). Soft flexible gripper design, characterization and application," in *Proceedings of the International Conference SCIT, Warsaw, Poland* (Berlin: Springer).

4

Antagonistic Actuation Principle for a Silicone-based Soft Manipulator

Ali Shiva[1], Agostino Stilli[1], Yohan Noh[1], Angela Faragasso[1], Iris De Falco[2], Giada Gerboni[2], Matteo Cianchetti[2], Arianna Menciassi[2], Kaspar Althoefer[1], and Helge A. Wurdemann[1]

[1]Department of Informatics, King's College London, London, United Kingdom
[2]The BioRobotics Institute, Scuola Superiore Sant'Anna, Pontedera (PI), Italy

Abstract

This chapter proposes an alternative actuation principle that investigates the capability of variable stiffness of a continuum silicon-based manipulator which was primarily developed for Minimally Invasive Surgery (MIS). Inspired by biological muscular composition, we have designed a hybrid actuation mechanism that can be alternatively used for STIFF-FLOP manipulator. The current soft robot is actuated by pneumatic pressure, in addition to incorporating tendons' tension, which are placed within the soft robot's body. Experiments are conducted by exerting an externally applied force in different poses, and simultaneously varying the stiffness via the tendons. Test results are demonstrated, and it is observed that dual antagonistic actuation, with the benefit of higher force capacities, could indeed promise enhancing soft robotics morphological features.

4.1 Introduction

Researchers have shown increasing interest in robotic systems which could potentially overcome some limitations of conventional robots with rigid joints and links [1]. Looking into biological examples, animals' appendages such as the octopus arm have been the inspiration for developing soft and hyper-redundant robots and aiming to achieve similar capabilities as their biological

65

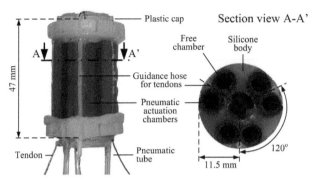

Figure 4.1   Side and cross-section view of a segment/module of the STIFF-FLOP manipulator with integrated stiffening mechanism based on the antagonistic principle: Three pairs of pneumatically actuated chambers are located in a silicone body. Between each set, a hose is integrated into the periphery of the manipulator to guide the tendons that are used to apply stiffening.

counterparts [2–5]. The application of these type of robots can result in significant improvements within a number of fields where traditional robots are currently used [6–8]. One of these areas is Minimally Invasive Surgery (MIS)—also called laparoscopic or keyhole surgery [9, 10]. In minimally invasive procedures, rigid laparoscopic tools are inserted through 12–15 mm incisions called Trocar ports which facilitates surgeries within the body [11]. Multiple challenges have been reported on the limited maneuverability of rigid surgical tools [11, 12] during a number of procedural steps in MIS such as posterior and lateral Total Mesorectal Excision (TME). In this regard, soft robotics has demonstrated immense potential [11]. The soft structure is beneficial due to reducing unwanted abrasion of internal tissue. The large Degrees of Freedom (DoF) shows promising for navigation around internal organs to reach a specific target, rather than cutting and therefore damaging healthy tissue. However, in employing soft robots, there exists a challenge on how to achieve higher stiffness and how to adjust it when dealing with different environments and tissues when we need to apply various forces [13].

A background on stiffness mechanisms is presented in Section 4.2. Section 4.3 gives the details of the dual actuation mechanism. The mechanical design of the soft, stiffness-controllable robot arm is presented in Section 4.4 along with the overall control architecture. Section 4.5 introduces the experimental methodology to investigate the efficiency of the variable stiffness mechanism and demonstrate the results. Finally, conclusions and future works are presented in Section 4.6.

4.2 Background

Recently, researchers have shown interest in finding appropriate methods for the position, force, and stiffness control of soft manipulators. Along these lines, the STIFF-FLOP focuses on studying such mechanisms within the octopus, and attempts to replicate some relevant biological characteristics to develop medical robotics systems for Minimally Invasive Surgery (MIS) [14] with integrated sensors [15–18]. Stiffness control is achieved with an embedded chamber within the silicone body containing granule that can be jammed by applying a vacuum [13, 19–21]. Hence, the manipulator's pose can be made more robust and resilient in a desired position. The concept of polymeric artificial muscles described in [22] to actuate a robot manipulator was furthered in [23] by integrating granule-filled chambers which when exposed to varying degrees of vacuum could actuate, soften, and stiffen the manipulator's joints. A similar concept is proposed in [24]. A hollow snake-like manipulator consists of multiple overlapping layers of thin Mylar film. By applying vacuum pressure, the friction between the film layers increases, which results in a tunable stiffness capability. In [25], the authors report on a thermally tunable composite for mechanical structures. This flexible open-cell foam coated in wax can change stiffness, strength, and volume. Altering between a stiff and soft state and vice versa introduces a time delay as the material does not instantly react to the heating-up or cooling-down process.

Here, the hybrid actuation principle has been applied to the STIFF-FLOP manipulator. Air pressure is mainly used for stretching out and controlling the motion and direction of the soft manipulator. When stiffness increase is required, tendons are manipulated in such a way as to oppose the pneumatic actuation. The results show the capabilities of adopting this antagonistic actuation scheme and the main advantages of the proposed technique when compared to traditional, single-actuation-type robot manipulators and to stiffening mechanism such as granular jamming.

4.3 Bio-Inspiration and Contributions

The work presented here has been inspired by the soft tentacles of octopi that demonstrate infinite DoFs. The tentacle comprises longitudinal and transverse muscles, bonded within the arm. Figure 4.2 gives an overview of the arm's structure. Different muscles are manipulated in such a way as to control the stiffness of its arm according to the nature of the task at hand.

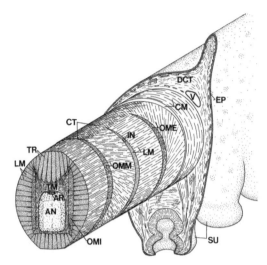

Figure 4.2   Three-dimensional illustration of an octopus tentacle with showing longitudinal muscle fibres (LM), transverse muscle fibres (TM) and connective tissue (CT). These types of muscles are primarily used by the octopus to create a range of stiffness along its arms [26].

Hence, by keeping the principles of operation of the biological counterpart in mind, our proposed manipulator utilizes the two actuation mechanisms: fluidic, and tendon-based, with the ability to oppose each other and thus capable of varying the arms' stiffness over a wide range. Hence, the proposed antagonistic actuation method bring together the advantages of fluidic and tendon-driven actuation. Tendon-based actuation is desirable when more precise position control is needed, and/or higher external forces are to be experienced. Stepper motors that are used to extend/flex each tendon are externally based [27, 28]. Pneumatic actuation is suitable for inherently safe scenarios such as handling delicate tasks where softness is required, in addition to generating a wider range of motion.

4.4 Integration of the Antagonistic Stiffening Mechanism

One segment/module of the STIFF-FLOP manipulator (see Figure 4.3) is a hollow cylinder of silicone made of Ecoflex® 00−50 Supersoft Silicone with material properties as shown in Table 4.1. The segment has an overall length of 47 mm and an outer diameter of 23 mm. In the periphery of this cylinder, three pairs of reinforced fluidic pressure chambers (6 mm diameter) are

Table 4.1 Technical properties of Ecoflex® 00 − 50 Supersoft Silicone[1]

Shore Hardness	Tensile Strength	Elongation at Break
00 − 50	315 psi	980%

implemented which are actuated pneumatically. Each dual fluidic chamber is connected to one 2 mm outer diameter inlet air pipe creating the ability to bend the module by increasing the air pressure in one dual chamber relative to the other two fluidic chambers. Simultaneous pressurization of the all dual chambers results in overall elongation of the segment. For the sake of completeness, the segment's structure in Figure 4.3 shows an inner free chamber of 9 mm diameter. This space is generated to pass through tubes from additional segments and wires from integrated sensory systems when creating a manipulator with a series of multiple modules.

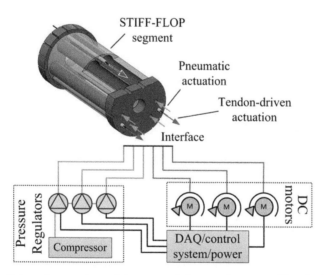

Figure 4.3 Schematic overview of the antagonistic actuation setup: The air chambers are connected to three pressure regulators. An air compressor supplies pressurised air to the regulators. Each tendon is wound around a pulley which is fixed to the shaft of a stepper motor. The analogue input for the three motors and three pressure regulator is controlled via a data acquisition board.

---

[1]Smooth-On, Inc. *Ecoflex® Series* Available on http://www.smooth-on.com/tb/files/ ECOFLEX_SERIES_TB.pdf, Accessed on May 2015.

4.4.1 Embedding Tendon-driven Actuation into a STIFF-FLOP Segment

The tendon-driven actuation mechanism is embedded into a single STIFF-FLOP segment. Figure 4.3 shows a side and cross-sectional view of the robot arm with the integrated antagonistic actuation principle. In this prototype, a hose (made of stretchable latex tubes with a diameter of 1.6 mm) is aligned in between each set of the fluidic chambers, hence being parallel to the longitudinal axis. The three hoses are placed 120° from each other feeding through tendons for extrinsic actuation. This design will allow the tendons to slide within the latex tubes and avoid any cuts into the silicone body. Due to the tube's material properties, the STIFF-FLOP segment keeps its key characteristics of being soft and squeezable; the latex tubes move in a compliant way when intrinsically actuating the robot.

Through each of the three passages, we pass a microfilament-braided PowerPro Super Line of 0.15 mm diameter acting as tendons. The three tendons are fixed to a plastic cap at the tip of the robot arm to distribute forces onto the soft tip surface when applying tension. The entire combination of the aforementioned configuration constructs one segment, which is shown in Figure 4.3.

4.4.2 Setup of the Antagonistic Actuation Architecture

The overall actuation architecture consists of an air compressor, three pressure regulators, a data acquisition board (DAQ), three stepper motors, and a modified STIFF-FLOP segment as described in Section 4.4.1. Figure 4.3 presents the interconnection within the test setup.

As mentioned earlier, a hybrid actuation mechanism is employed here: On the pneumatic actuation side, an air compressor (BAMBI MD Range Model 150/500) supplies the required pressurised air of 5 bar to three independent pressure regulators (SMC ITV0030-3BS-Q). The output three dual fluidic chambers of the soft module via with 0 to 10 VDC input signal and a set pressure range of 0.001 to 0.5 MPa. Each pressure regulator adjusts the outlet pressure for each dual fluidic chamber according to the command received from the computer through a DAQ board (NI USB-6411).

On the tendon side, each tendon is connected to a stepper motor (Changzhou Songyang Machinery & Electronics Co. SY57ST56-0606B) which provides a holding torque of 0.59 Nm. Each stepper motor has a pulley attached on the output shaft which the tendon is wound around. The pulley has 6.4 mm radius, which results in a maximum of 92.6 N of tension.

Since one STIFF-FLOP segment has three tendons, three stepper motors are taken into consideration. Each stepper motor is driven via a driver (Big Easy Driver ROB-11876), which communicates with the computer via a DAQ board. The computer has a Windows-based operating system with the related programming in C++.

4.5 Test Protocol, Experimental Results, and Discussion

4.5.1 Methodology

Various experiments for stiffness characterization have been carried out by hanging the manipulator and exerting external force at the distal end. A command for the computer generates a displacement of 1 cm for the linear rail mechanism which presses against the tip of the module. Reaction forces are recorded utilizing a Nano17 Force/Torque sensor by ATI Industrial Automation. The experimental configurations were defined as follows:

Scenario 1:
The module is held vertically downwards. The force is applied laterally to the tip as shown in Figure 4.4(a). In this scenario, four different sub-cases are investigated:

A No air pressure and no tendon tension.
B Equally air-pressurised chambers (i.e. elongation) with no tendon tension.
C No air pressure with initial equal tendon tension.
D Equally air-pressurised chambers with initial equal tension in tendons.

Scenario 2:
The module is held vertically and one of the dual chambers is pressurised to form a 90% curved shape, and the force is applied laterally as shown in Figure 4.4(b). Two different sub-cases are investigated:

A One pressurized chamber and no tendon tension.
B One pressurized chamber and tension in the tendons.

Scenario 3:
The module is pressurized to be configured as Scenario 2. However, the force is applied opposing the tip as shown in Figure 4.4(c). Also in this scenario, two different sub-cases are investigated:

A One pressurised chamber and no tendon tension.
B One pressurized chamber and tension in the tendons.

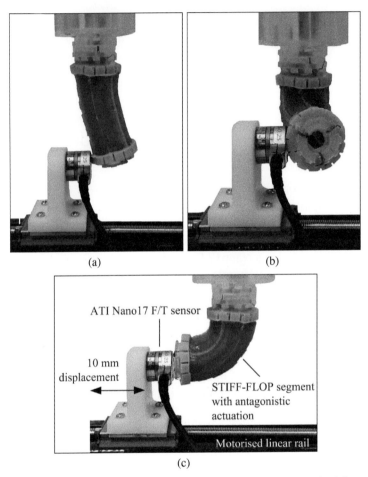

(a)   (b)

(c)

Figure 4.4   An ATI Nano17 Force/Torque sensor is mounted on a motorised linear mechanism displacing the manipulator's tip by 1 cm: The configurations in (a), (b), and (c) show Scenarios 1, 2, and 3, respectively.

4.5.2 Experimental Results

Data from the ATI Nano17 F/T sensor and corresponding displacement of the motorized linear module were recorded at 1 kHz using a DAQ card (NI USB-6211). Four trials were performed for each sub-case.

Experimental results of all four sub-cases of Scenario 1 are presented in Figure 4.5(a). When the module is neither pressurized nor stiffened by tendons, the amount of its resistive force subjected to a 1 cm lateral displacement

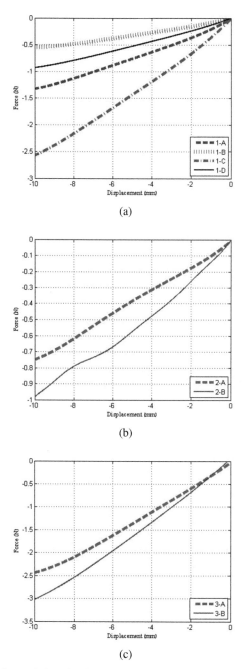

Figure 4.5 Experimental data for Scenarios 1, 2, and 3. Forces have been recorded for displacements of 1 cm of the manipulator's tip. Table 4.2 summarises the data analysis.

is about 1.32 N. This value is 0.55 N when all three chambers are pressurized. When subjected to tendon stiffening, the resistive forces displayed by the module reach to 2.56 N and 0.93 N, respectively, showing a 94% and 69% increase compared to the first and second sub-case.

The results of the two sub-cases of Scenario 2 are shown in Figure 4.5(b). When the module is only pressurised, the value of the resistive force is 0.75 N. With tendon stiffening is added to the module, this resistive force increases to 0.98 N showing a 31% growth.

The results of the two sub-cases for Scenario 3 are presented in Figure 4.5(c). It can be seen that in the presence of pressure only, the module generates a resistive force of 2.43 N. However, by introducing tendon stiffening, the resistive force due to 1 cm displacement intensifies to 3.02 N, displaying a 24% growth.

Table 4.2 summarizes the experimental results. For each sub-case, the maximum force, hysteresis, and percentage of increase is calculated.

4.5.3 Discussion

Experimental results in Table 4.2 and Figure 4.4 demonstrate that incorporating the dual actuation mechanism has improved the stiffness of the STIFF-FLOP module up to almost 100%. In this sense, we are able to deal with higher external forces, and/or resist external disturbances more effectively during task execution. This gives the surgeon the ability to move the manipulator about primarily with pressure actuation, and thereafter, use the tendon stiffening to acquire not only higher stiffness values, but also fine-tune the final position of the end effector for more accurate maneuvering.

Table 4.2   Summarized results of stiffness tests for Scenarios 1, 2, & 3

	Scenarios				F_{max}	Hyst.	Increase
1-A	Tens.	No	Press.	No	1.32 N	21.6%	n/a
1-B	Tens.	No	Press.	Yes	0.55 N	27.2%	n/a
1-C	Tens.	Yes	Press.	No	2.56 N	18.9%	93.9%
1-D	Tens.	Yes	Press.	Yes	0.93 N	28.5%	69.1%
2-A	Tens.	No	Press.	Yes	0.75 N	21.8%	n/a
2-B	Tens.	Yes	Press.	Yes	0.98 N	33.46%	30.7%
3-A	Tens.	No	Press.	Yes	2.43 N	27.47%	n/a
3-B	Tens.	Yes	Press.	Yes	3.02 N	14.86%	24.3%

Table 4.3 Force results for granular jamming applying a 10 mm displacement as reported in [9]

Scenarios	Granular Jamming	F_{max}	Increase
1-A	Off	2.2 N	n/a
1-A	On	3.1 N	40.9%
2-A	Off	2.3 N	n/a
2-A	On	2.7 N	17.4%
3-A	Off	2.8 N	n/a
3-A	On	3.3 N	17.9%

In [9], an 8 mm diameter channel of granular material, coffee, was embedded into a previous version of the silicone-based STIFF-FLOP module. The length of this segment was 50 mm with the silicone structure having a diameter of 25 mm. The pneumatically actuated chambers were not reinforced; a crimped, braided sheath of 35 mm covered the silicone structure and prevented a ballooning effect. Neglecting the outer cover, the STIFF-FLOP module has similar dimensions as the segment used in this chapter. The key experimental results for stiffness tests at a displacement of 10 mm are summarized in Table 4.3. The test configurations of three scenarios are equivalent to the ones described in Section 4.5.1; however, stiffening is activated by applying a vacuum of 36%.

Comparing Tables 4.2 and 4.3, the actual maximum forces F_{max} measured during the experimental tests of Scenarios 1 and 2 are larger using granular jamming. The presence of coffee granules (under atmospheric or vacuum pressure) integrated into the silicone-based robot results in a stiffer module. Looking at the percentage increase caused by granular jamming on the one hand and the antagonistic mechanism on the other hand, the bio-inspired tendon-based stiffening principle is able to generate a larger growth. Additionally, adding the tendons has not only resulted in the ability to vary the stiffness, but also acts beneficially towards the position control of the soft manipulator allowing simultaneous position and stiffness control.

4.6 Conclusions

In this chapter, we have proposed the dual actuation mechanism to a soft continuum robot, which enables antagonistic stiffening. This approach can be used as an alternative to the actuation method proposed in Chapter 3. Hereby, air pressure is employed for a wider range of movement such as elongation and bending, while the tendon-driven system provides the fine-tuned movements in addition to the ability of varying the stiffness. Due to the internal accommodation of the tendons in the wall of the module, this added feature does not increase the overall diameter nor the wall thickness of this soft continuum robot.

Future work is to incorporate a system which would allow adjustment of tendon tensions in a more ergonomic manner.

4.7 Funding

The work described in this chapter is partially funded by the Seventh Framework Programme of the European Commission under grant agreement 287728 in the framework of EU project STIFF-FLOP.

References

[1] Ataollahi, A., Karim, R., Fallah, A. S., Rhode, K., Razavi, R., Seneviratne, L. T., et al. (2013). 3-DOF MR-compatible multi-segment cardiac catheter steering mechanism. *IEEE Trans. Biomed. Eng.* 99:1.

[2] Buckingham, R. (2002). Snake arm robots. *Ind. Robot* 29, 242–245.

[3] Caldwell, D. G., Medrano-Cerda, G. A., and Goodwin, M. (1995). "Control of pneumatic muscle actuators," in *Proceedings of the IEEE Control Systems*, San Diego, CA, 15.

[4] Cheng, N. G., Gopinath, A., Wang, L., Iagnemma, K., and Hosoi, A. E. (2014). Thermally tunable, self-healing composites for soft robotic applications. *Macromol. Mater. Eng.* 299, 1279–1284.

[5] Cianchetti, M., Arienti, A., Follador, M., Mazzolai, B., Dario, P., and Laschi, C. (2011). Design concept and validation of a robotic arm inspired by the octopus. *Mater. Sci. Eng. C* 31, 1230–1239.

[6] Cianchetti, M., Ranzani, T., Gerboni, G., de Falco, I., Laschi, C., and Menciassi, A. (2013). "STIFF-FLOP surgical manipulator: mechanical design and experimental charaterization of the single module," in *Proceedings of the IEEE/RSJ International Conference on Intelligent Robots and Systems*, Hamburg.

[7] Cianchetti, M., Ranzani, T., Gerboni, G., Nanayakkara, T., Althoefer, K., Dasgupta, P., et al. (2014). Soft robotics technologies to address shortcomings in today's minimally invasive surgery: The STIFF-FLOP approach. *Soft Robotics* 1, 122–131.

[8] Cieslak, R., and Morecki, A. (1999). Elephant trunk type elastic manipulator—a tool for bulk and liquid type materials transportation. *Robotica* 17, 11–16.

[9] Fras, J., Czarnowski, J., Macias, M., Glowka, J., Cianchetti, M., and Menciassi, A. (2015). "New STIFF-FLOP module construction idea for improved actuation and sensing," in *Proceedings of the ICRA*, New Orleans, LA.

[10] Godage, I. S., Nanayakkara, T., and Caldwell, D. G. (2012). "Locomotion with continuum limbs," in *Proceedings of the IEEE/RSJ International Conference on Intelligent Robots and Systems*, Deajeon.

[11] Gravagne, I., Rahn, C., and Walker, I. D. (2003). Large deflection dynamics and control for planar continuum robots. *IEEE/ASME Trans. Mech.* 8, 299–307.

[12] Jiang, A., Aste, T., Dasgupta, P., Althoefer, K., and Nanayakkara, T. (2013). "Granular jamming with hydraulc control," in *Proceedings of the ASME International Design Engineering Technical Conferences and Computers and Information in Engineering Conference*, New York, NY.

[13] Jiang, A., Secco, E. L., Wurdemann, H., Nanayakkara, T., Dasgupta, P., and Athoefer, K. (2013). "Stiffness-controllable octopus-like robot arm for minimally invasive surgery," in *Proceedings of the 3rd Joint Workshop on New Technologies for Computer/Robot Assisted Surgery*, Pisa.

[14] Jiang, A., Secco, E., Wurdemann, H. A., Nanayakkara, T., Dasgupta, P., and Athoefer. K. (2013). "Stiffness-controllable octopus-like robot arm for minimally invasive surgery," in *Proceedings of the Workshop on New Technologies for Computer/Robot Assisted Surgery*, Pisa.

[15] Jiang, A., Xynogalas, G., Dasgupta, P., Althoefer, K. and Nanayakkara, T. (2012). "Design of a variable stiffness flexible manipulator with composite granular jamming and membrane coupling," in Proceedings of the *IEEE/RSJ International Conference on Intelligent Robots and Systems*, Vancouver.

[16] Kier, M. W., and Stella, M. P. (2007). The arrangement and function of octopus arm musculature and connective tissue. *J. Morphol.* 268, 831–843.

[17] Kim, Y. J., Cheng, S., Kim, S., and Iagnemma, K. (2012). "Design of a tubular snake-like manipulator with stiffening capability by layer jamming," in *Proceedings of the IEEE/RSJ International Conference on Intelligent Robots and Systems*, Deajeon.

[18] Laschi, C., Mazzolai, B., Mattoli, V., Cianchetti, M., and Dario, P. (2009). Design of a biomimetic robotic octopus arm. *Bio. Biomimetics* 4, 1–8.

[19] Li, M., Ranzani, T., Sareh, S., Seneviratne, L. D., Dasgupta, P., Wurdemann, H. A., and Althoefer, K. (2014). Multi-fingered haptic palpation utilising granular jamming stiffness feedback actuators. *Smart Mater. Struct.* 23:095007.

[20] McMahan, W., Jones, B. A., and Walker, I. D. (2005). "Design and implementation of a multi-section continuum robot: Air-octor," in *Proceedings of the IEEE/RJS International Conference on Intelligent Robots and Systems*, Deajeon.

[21] Neppalli, S., Jones, B., McMahan, W., Chitrakaran, V., Walker, I., Pritts, M., et al. (2007). "Octarm—a soft robotic manipulator," in *Proceedings of the IROS*, Hamburg.

[22] Noh, Y., Sareh, S., Back, J., Wurdemann, H. A., Ranzani, T., Secco, E. L., et al. (2014). "A three-axial body force sensor for flexible manipulators," in *IEEE International Conference on Robotics and Automation*, Montreal.

[23] Noh, Y., Secco, E. L., Sareh, S., Wurdemann, H. A., Faragasso, A., Back, J., et al. (2014). "A continuum body force sensor designed for flexible surgical robotic devices," in *IEEE Engineering in Medicine and Biology Society*, Stockholm.

[24] Ranzani, T., Gerboni, G., Cianchetti, M., and Menciassi, A. (2015). A bioinspired soft manipulator for minimally invasive surgery. *Bio. Biomimet.* 10:035008.

[25] Sareh, S., Jiang, A., Faragasso, A., Noh, Y., Nanayakkara, T., Dasgupta, P., et al. (2014). "Bio-inspired tactile sensor sleeve for surgical soft manipulators," in *Proceedings of the ICRA*, Stockholm.

[26] Walker, I. D., Dawson, D. M., Flash, T., Grasso, F. W., Hanlon, R. T., Hochner, B., et al. (2005). "Continuum robot arms inspired by cephalopods," in *Proceedings of the UGVT VII*, Orlando, FL.

[27] Xie, H., Jiang, A., Wurdemann, H. A., Liu, H. Seneviratne, L. D. and Althoefer, K. (2014). Magnetic resonance-compatible tactile force sensor using fibre optics and vision sensor. *IEEE Sensors J.* 14, 829–838.

[28] Yamashita, H., Iimura, A., Aoki, E., Suzuki, T., Nakazawa, T., Kobayashi, E., et al. (2005). "Development of endoscopic forceps manipulator using multi-slider linkage mechanisms," in *Proceedings of the 1st Asian Symposium on Computer Aided Surgery—Robotic and Image guided Surgery*, Ibaraki.

5

Smart Hydrogel for Stiffness Controllable Continuum Manipulators: A Conceptual Design

Daniel Guevara Mosquera[1], S.M. Hadi Sadati[1,3], Kaspar Althoefer[2] and Thrishantha Nanayakkara[3]

[1]Center for Robotic Research (CoRe), King's College London, London, United Kingdom
[2]School of Engineering and Materials Science, Queen Mary University of London, London, United Kingdom
[3]Dyson School of Design Engineering, Imperial College London, London, United Kingdom

Abstract

As it was discussed in previous chapters of Part I, soft continuum trunk and tentacle manipulators have high inherent dexterity and reconfigurability and have become an attractive candidate for safe manipulation and explorations in surgical and space robotic applications, recently. However, achieving accuracy in precise tasks is a challenge with these highly flexible structures, for which stiffness variable designs based on jamming, smart material, antagonistic actuation, and morphing structures have been introduced in the recent years. In this chapter, variable stiffness properties of an electro-active poly (sodium acrylate) (pNaAc) hydrogel are tested. An anisotropic stiffness ion pattern is printed on the hydrogel straps giving them shape memory properties. The hydrogel swells up to two times its original size and soften (4.2-0 KN/m) in an ethanol aquatic solution depending on the ethanol saturation while preserving its programmed shape. Changing the solution ethanol saturation can be used to control the hydrogel stiffness based of which a conceptual design for a stiffness controllable STIFF-FLOP module is presented and will be fabricated in the future.

5.1 Introduction

Performing complicated biological tasks such as manipulation in unpredictable conditions where safe interaction with the environment is important requires dexterity and compliance in which low actuation energy, high dexterity, reachability, maneuverability, back drivability, and self adjustablbility of continuum mechanisms are shown to be advantageous [1]. In nature, biological creatures benefit from compliant muscle-tendon-bone structures capable of exerting instantaneous high-peak forces and velocities necessary for such tasks [2]. On the other hand, to address the common problems with current actuation methods in robotics research, such as back-drivability, stiffness control, and energy consumption, different methods such as compliance actuation [3], antagonistic actuation [4, 5], reconfigurable design [6], and more recently the use of stiffness tuneable material [7], and morphing structures [8] are employed. The control of damping to achieve a desired stiffness is shown to be important too, for task accuracy and control stability [9, 10]. However, compliance has disadvantages such as reduced control bandwidth, stability issues, and underdamped modes where high stiffness modes are required to achieve precision in tasks involving working against external loads [10].

Soft continuum manipulators, mostly inspired by octopus arms, snake, land animals' tongue, and trunk, with high inherent dexterity and reconfigurability, have become an attractive candidate for safe manipulation and explorations in surgical and space robotic applications in recent years. The passive shape adaptation and large reachable configuration space features of this class of manipulators due to their highly deformable nature made them a perfect choice for minimally invasive insertion of surgical tools in the confined maze-like space in a robotic surgery [11, 12]. Among the continuum manipulator designs, braided pneumatic and hydraulic actuators provide a uniform homogeneous deformation, robust geometry and force control, and linear and reversible behavior [13–16] compared to the non-braided versions [17–19] which circumferential expansion limits their application in confined space. Besides, accuracy in precise tasks is a challenge with highly flexible structures for which stiffness variable designs based on jamming, smart material, antagonistic actuation, and morphing structures have been introduced in the recent years [20]. The uncertainty in the material deformation due to highly elastic environment, insufficient flexibility, and lack of control feedback are the limitations of continuum manipulators [21]. While most of the research have been focused on the design and modeling of soft manipulators, methods of stiffness control for soft media have recently shown to be important for efficient minimalist actuation, minimal invasive interaction, and control and sensing precision [22].

Jamming concept and stiffness tunable material are used in variable compliance robotic studies [20]. Jamming concept, for stiffness control through modulating the Coulomb friction and viscous damping between the jammed media, has been utilized in the design of stiffness controllable actuators [23], flexible manipulators [24], variable stiffness joints [25], rehabilitation devices [26], stiffness displays [27], biomimetic organs [28], reconfigurable mechanisms [29], and grippers [30] of which a comprehensive review is presented in [20]. The comparative study by Wall et al. on granular, layer, and scale jamming used in a selective stiffness controllable pneumatic actuator, PneuFlex, showed the layer jamming arrangement to be the most capable design with an eight times increase in the stiffens and 2.23 times increase in the resisting force [23]. Tendon driven jamming was introduced recently to overcome portability limitation of pneumatic enabled jamming in underwater and space applications where a continuum rod flexural stiffness is controlled by modulating the shear friction force between the scales [28, 31] and helical rings [32]. The resulting Coulomb damping opposes the inter-layer shear forces caused by the external load bending momentum.

Thermo-active stiffness tuneable structures and materials have attracted many research in the past few years due to their high range of stiffness change, easy electrical modulation through heating, and possible 3D printing fabrication [33]. Low melting point (LMP) alloy such as field's metal [33–39] and LMP composite material with inherent thermal instability such as wax [40] and ABS which are used as the base material in many standard 3D printing devices have been utilized to design 3D printable thermally stiffness tunable structures [41–44]. As a result, new compound micro actuators capable of actuation and shape fixation with high reversibility and load bearing capacity are designed such as a thermally stiffness tunable actuator fiber in [45] and a thermally tunable shape memory electro-active polymer in [46].

Granular [16] and tendon [47] stiffening methods were investigated for a STIFF-FLOP module recently. Furthermore, we presented a tendon driven scale jamming design inspired by the helical arrangement and morphology of the fish scales (Figures 5.1a and b) to control torsional stiffness of a helical interface cross section for STIFF-FLOP manipulators in our previous work. As a result, we achieved a simpler design and actuation method which, for the first time, provides better wearability, higher stiffening ratio, linear behavior, longer axial stretch, and lower hysteresis [31] compared to the

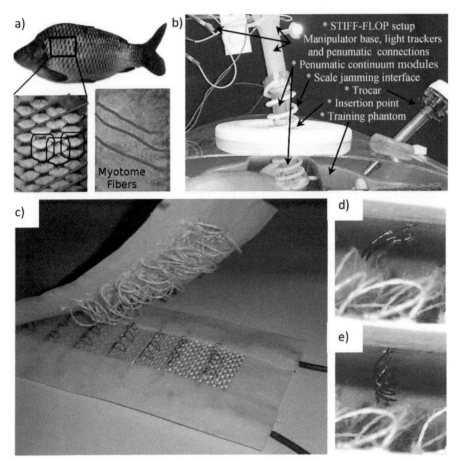

Figure 5.1   Bio-inspired stiffening mechanisms developed for STIFF-FLOP continuum actuators: a scale jamming interface inspired by helical myotome attachment sites and overlapping
scales in a Cyprinus carpio fish (a), sample application of the bio-inspired scale jamming
interface on a STIFF-FLOP continuum actuator (b) [31], a wearable electro-active Velcro
attachment mechanism [48] inspired by stiffening mechanism in natural wood using an inter-
layer micro hook structure [49] (c), and shape memory alloy hooks' in relaxed (d) and
activated (e) states [48].

previous stiffenong solutions for STIFF-FLOP [16, 47] and, to the best of
our knowledge, other available locking and jamming designs in literature
[20]. However, local and directional stiffness control of continuum manipulators are still challenging. While usually the normal forces on the jammed
surfaces are controlled for stiffening, in our recent work, we investigated

the idea of using an active attachment mechanisms between the layers based on a novel electro-active velcro using shape memory alloy wire [48] (Figures 5.1c to d).

As the most similar artificial product to a biological tissue, hydrogels are organic soft materials made of crosslinked polymers capable of absorbing great amount of water without losing their initial structure integrity. These chemical crosslinks provide high mechanical strength and long degradation time and have become widely studied as soft actuators [50], micro robots for drug delivery [51], tissue engineering [52], biosensor [53], self-healing structures [54], etc. Active hydrogels show stiffness controllable properties. Most environmentally responsive hydrogels consist of monomers, an initiator, and an accelerator [53, 55]. The fabrication process usually consists of free radical polymerization of low-molecular-weight monomers in aqueous solution with a crosslinking agent. The use of a crosslinker results in a gel that expands due to the monomer/crosslinker concentrations and the polymerization conditions. UV or thermal treatment can be used for the polymerization process based on the selected solvent. The polymerization technique changes the formed gel properties [56, 57].

In this research, we investigate the fabrication and use of an electro-active stiffness tunable poly(sodium acrylate) (pNaAc) hydrogel as presented in [57] with possible application in stiffness controllable continuum manipulators. We focused on the synthesis, characterization, preparation, and possible applications of pNaAc hydrogel. Different properties of an easy to fabricate hydrogel as in [57] are investigated. A conceptual design for a stiffness controllable STIFF-FLOP module is presented based on a porous hydrogel shell with an anisotropic stiffness ion pattern. The use of active hydrogels for stiffness control of continuum mechanisms is a novel concept which our results in this section provide the preliminary understanding about the gel behavior and a proof for such concept. Our approach in using conventional set of tools to fabricate an active hydrogel shows the feasibility of fabrication of such smart material to be used by researchers with limited to no expertise in experimental chemistry.

Materials and methods of our research are presented in Section 5.2, where fabrication of two slightly different pNaAc hydrogels is discussed. Experiments on the samples' swelling and stiffness properties and a conceptual design for a porous pNaAc hydrogel shell for continuum manipulators are discussed in Section 5.3. Finally, conclusions are presented in Section 5.4

5.2 Materials and Methods

Active hydrogels usually consist of monomers, an initiator, and an accelerator [55]. Ammonium persulfate (APS) is usually used as an initiator and tetramethylethylenediamine (TEMED) as a catalyst. The persulfate can be used to convert monomers to free radicals. The free radicals react with unactivated monomers and begin the polymerization process. The APS (initiator) and TEMED (accelerator) are added to an aqueous solution of albumin [58], polyacrylate [59], polyvynil alcohol [60], or N,N-methylenebis(acrylamide) [57] based on the required hydrogel properties.

Among the different recent fabrication methods introduced in the literature, we used an easy method as in [57], with some changes in the mixture and mole ratios, to fabricate an electro- and thermo-active hydrogel, cable of generating mechanical motion and stiffness variation in dry and wet conditions. Apart from the necessary chemical compounds and some conventional tools for measuring and mixing, this method needs only a conventional low power (70 W) nail polishing UV dryer and a conventional kitchen oven for polymerization and drying processes. Acrylamide (AAm, Sigma: www.sigmaaldrich.com), anhydrous acrylic acid 99% (AA, Sigma), poly(N-isopropylacrylamide) (pNIPAAm, Sigma), N,N-methylenebis(acrylamide) (Sigma), ammonium persulfate (APS, Sigma), NaCl, N,N,NN-tetramethylethylenediamine 99.5% (TEMED, Sigma), agarose LE (Sigma), sodium hydroxide (Fisher Scientific: www.fishersci.co.uk), ethylenediaminetetraacetic acid (EDTA, Sigma), and Milli-Q deionized water (18.2 MΩ cm, Amazon) are used for the fabrication of the hydrogel straps as in [57]. Copper anode wires (Cu^{2+}, 1 mm diameter, Alfa Aesar: www.alfa.com), ethanol, and deionized distilled water (Amazon) are used for providing aquatic or dry test environment, electric activation, and material recovery.

5.2.1 Active Hydrogel Preparation

Following the procedure in [57], we fabricated two poly(sodium acrylate) (pNaAc) hydrogel samples with slightly different compound ratios considering our stiffness controllable application. The first strap sample, to be formed in a helical shape after ion pattern printing (Figure 5.2), is prepared by free radical polymerization in an aqueous solution, combining 3 mole (M) concentration of poly(N-isopropylacrylamide) (pNIPAAm) monomer, as a stimuli responsive substance, and N,N-methylenebis(acrylamide) as the crosslinker with mole ratio (divinyl to vinyl monomers) of 1:100. It was

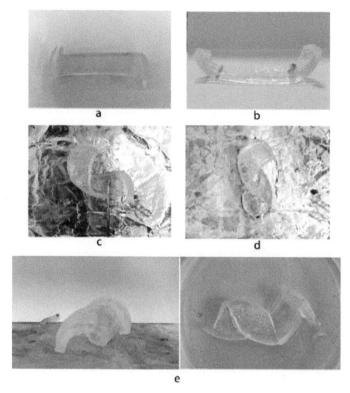

Figure 5.2 A sample electro-active swelling hydrogel strap with 1 mm thickness (a), printed ionic patterns with light blue color on the two sides of the strap (b), printing ionic patterns with a copper wire (Cu^{2+}) (c), helix formation of the strap due to mechanical shrinkage at the printed ionic pattern (d), and side and top views of the final helical hydrogel sample when it is still wet (e).

mixed with fluorescein isothiocyanate (FITC 5 m diameter) for electrostatic stabilization. As a result, a gel with a large swelling and stiffness variation response in different aquatic solutions is fabricated. It was prepared with Milli-Q deionized water (18.2 MΩ cm) and ethanol solution with 4:1 ratio. The solution was subjected to 70 W UV light for 1 hour and then dried at 60°C temperature in an oven for free radical polymerization process.

The second strap sample, to be formed in a cylindrical shape after ion pattern printing (Figure 5.3), is prepared by free radical polymerization of AAm monomer with a small amount of N,N-methylenebis(acrylamide) as

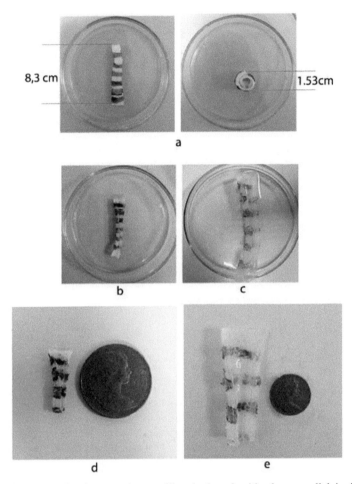

Figure 5.3 A sample electro-active swelling hydrogel with close parallel ionic printed patterns, forming a cylindrical shape, in top and side views (a), the sample after swelling due to immersing in 1:4 ethanol aquatic solution for 3 minutes (b) and in dry state after immersing in ethanol (c), the sample size in comparison to a British one pence coin in dry (d) and wet (e) conditions. The sample retains its cylindrical shape in dry and wet conditions regardless of 180% change in its size due to swelling and a softer structure in the swelled condition.

the crosslinker. Five grams of AA-AAm mixture is mixed with equimolar amounts of sodium hydroxide, based on the number of AA moles (mixture I); 240 mg of N,N-methylenebis(acrylamide) was added to 40 mg of APS which then was dissolved in distilled water and added to the mixture I (mixture II); 0.024 mL of TEMED was added to the mixture II and the product is molded in a thin layer shape. The equilibrium swelling ratio is achieved based on the crosslinker mole ratio (divinyl to vinyl monomers) which was set to be 5:100% here. Leaving the mixture for a whole day in the room ambient, the polymerization process produces a dry hydrogel that can be cut in the desired shape. The dry pNaAc hydrogel was immersed in four aquatic solutions, 1) distilled water and ethanol (EtOH), 2) distilled water, 3) deionized water, and 4) tap water for over 2 hours, to find the best swelling ratio, and then dried in 70°C using the oven. Watering and drying steps are repeated if the final mechanical properties are not satisfactory.

5.2.2 Active Hydrogel Properties and Ion Pattern Printing

The pNaAc hydrogels swell when are placed in an aquatic solution, usually for more than 3 minutes. The watering can be reversed, and the sample shrinks and hardens by putting it in ethanol for about 2 hours. The swelling deformation can be used as a mechanical actuation method. Besides, different electro-mechanical responses, e.g., swallowing, bending, and twisting can be achieved by ion printing, creating an anisotropic stiffness network consisting of conductive particles. This is done by applying 9-V DC through a copper wire (Cu^{2+}, anode) while placing the hydrogel straps on an aluminum foil (cathode). As a result, two adjacent sodium ions in the gel move toward a cupric ion and bind to the gel carboxylic groups. This leaves a ionized pattern with light blue color that shrinks and stiffens compared to the rest of the gel. The sample shape is fixed, showing a shape memory feature in Figures 5.2a to c. Figure 5.2 shows a sample with two ionic patterns at each end that bring the hydrogel strap to a helical shape and Figure 5.3a shows a sample with closer parallel patterns that forms a cylindrical shape. The samples maintain their shape while absorbing water solution as in Figures 5.2e and 5.3b,c. The inhomogeneous stiffness due to the ionic patterns causes an even more local shrinkage of the sample when immersed and dried in ethanol. The electric field deforms the hydrogel in air. This deformation is reversible by applying a reversed field. Both the local shrinkage due to the ionic actuation and watering-ethanol drying cycles of a ion-printed gel can be used as actuation mechanisms for a hydrogel actuator [57]. We investigated the watering-ethanol drying cycle here as a stiffening mechanism where ion patterns help bringing the structure to more stable and stiffer, e.g., helical or cylindrical, configurations. After multiple watering-ethanol drying cycles, we used the resulted dried contracted hydrogel in our stiffness tests.

5.3 Experiments and Discussion

5.3.1 Swelling Test

The dried hydrogels in ethanol swell up to 100% for the helical (Figure 5.2e) and 180% for the cylindrical (Figures 5.3b and c) shape strap after are immersed in deionized water for about 3 minutes (Figure 5.4) while maintaining but loosing the grip of its programmed shape. The sample can be dried in ethanol and immersed again with no noticable hysteresis buildup in our six trials. Leaving the swelled samples in the room temperature to dry disintegrates the hydrogel and breaks their structure as the water evaporates, while the dried sample in ethanol can be left for a long time unchanged in the room temperature, 1 week in our case, with no noticeable change in its swelling ability, final volume, and stiffness. The hardened gel in ethanol turns from a white transparent to a yellow cloudy color after being left in the room temperature for a while (Figure 5.5a). Despite the low stiffness of the swelled hydrogel in water, the dried samples in ethanol shrink and hold a stiff, hard to break, or deform shape, with a harder grip to the programmed shape. To achieve higher swelling ratios (lower stiffness), the sample needs to be fully dried first. Different aqueous solutions were tested, showing the best swelling results for deionized and common tap (with unknown chemical impurity) water while ethanol solution results in the least swelling ratio.

5.3.2 Stiffness Test

A helical pNaAc hydrogel strap, with size $55 \times 10 \times 1$ mm when swells, is prepared for stiffness test after initiating in 70 W UV light for 30 minutes,

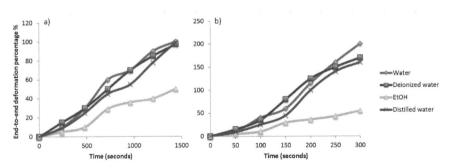

Figure 5.4 End-to-end deformation percentage (%) after swelling in different aquatic solutions for the pNaAc helical (Figure 5.2) (a) and cylindrical (Figure 5.3) (b) hydrogel samples.

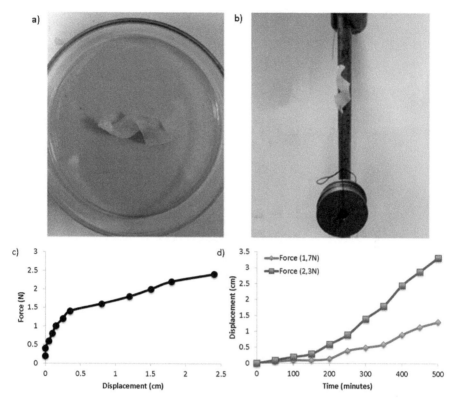

Figure 5.5 Hardened and dried electro-active hydrogel of Figure 5.2 in ethanol and room temperature (a), tension tests (b), force–displacement (c), and displacement–time (d) graphs for the simple tension experiments.

equilibrated in dionized water for 30 minutes, hardened in ethanol, and left in room temperature to completely dry for 3 days. The force–displacement and displacement–time graphs for the simple tension experiments are presented in Figures 5.5c and d, showing an initial quasi-static elastic deformation region with yield force of 1.5 N and stiffness of 4.2 KN/m followed by an accelerating linear plastic deformation region with resisting force up to 2.5 N and stiffness of 0.5 KN/m (Figure 5.5c). The structure breaks at 2.8 N. The hydrogel elasticity remains almost constant for 150 seconds. However, it decreases rapidly over time under a constant force (Figure 5.5d) which can be due to propagation of cracks in the gel structure. The hydrogel stiffness is negligibly low after placing it in an aquatic solution for 3 minutes. The large stiffness variation achieved by changing the ethanol saturation in an aquatic

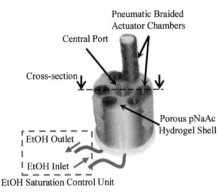

Figure 5.6   Conceptual design for a STIFF-FLOP module with a cylindrical porous pNaAc hydrogel with ion-printed ring patterns. The hydrogel stiffness is controlled by changing the ethanol saturation in EtOH solution, that fills the hydrogel shell pores, using a set of micro pumps and a solution saturation control unit.

solution can be exploited for stiffness variable applications such as stiffness controllable continuum robots and stiffness displays. A conceptual design is presented in Figure 5.6 where the body shell of a STIFF-FLOP manipulator is filled with a porous pNaAc hydrogel while the pores are filled with an aquatic ethanol solution. Changing the ethanol saturation by pumping in new solution with different ethanol saturation in the gel pores while draining out the previous solution results in a change in the structure stiffness. The slow stiffness variation is not a problem in medical tasks with mostly quasi-static motions. Printing ionic patterns, i.e. parallel rings on the structure, strengthen the structure by programming a cylindrical shape memory that can act as an external reinforcement for achieving even higher stiffness values.

5.4 Conclusion and Future Works

Electroactive hydrogels are usually formed based on monomers, an initiator, and an accelerator. Following the procedure presented in [57], we fabricated an electro- and thermo-active pNaAc hydrogel on which ionic patterns with anisotropic stiffness are printed using a copper wire as the anode and an aluminum sheet as the cathode. The ionic patterns shrink and stiffen bringing the gel to a programmed complex geometrical shape, helix and cylindrical in our case. The ionization process is reversible by reversing the electric field which can be used to design an electro-active hydrogel actuator. The hydrogel softens and swells up to two times of its original size in an aquatic solution while preserving its programmed shape. The observed stiffness variation (0–4.2 KN/m) by changing the ethanol saturation in an aquatic ethanol solution can be exploited to design stiffness controllable continuum manipulators or stiffness displays. A conceptual design for a stiffness controllable STIFF-FLOP module is presented with porous pNaAc hydrogel shell and ring-shaped ionic patterns, the pores of which are filled with aquatic ethanol solution. The structure stiffness can be controlled by changing the ethanol solution ratio using micropumps and a control unit to monitor and modulate the solution ratio. We plan to fabricate the proposed design in a future research.

References

[1] Albu-Schäffer, A., and Bicchi, A. (2016). "Actuators for Soft Robotics," in *Springer Handbook of Robotics*, eds B. Siciliano and O. Khatib (Cham: Springer International Publishing), 499–530.

[2] Tomori, H., Nagai, S., Majima, T., and Nakamura, T. (2013). "Variable impedance control with an artificial muscle manipulator using instantaneous force and MR brake," in *Proceeding of 2013 IEEE/RSJ International Conference on Intelligent Robots and Systems*, Tokyo, 5396–5403.

[3] Zhu, Y., Yang, J., Jin, H., Zang, X., and Zhao, J. (2014). "Design and evaluation of a parallel-series elastic actuator for lower limb exoskeletons," in *Proceeding of 2014 IEEE International Conference on Robotics and Automation (ICRA)*, Hong Kong, 1335–1340.

[4] Maghooa, F., Stilli, A., Noh, Y., Althoefer, K., and Wurdemann, H. A. (2015). "Tendon and pressure actuation for a bio-inspired manipulator based on an antagonistic principle," in *Proceedings of 2015 IEEE International Conference on Robotics and Automation (ICRA)*, Seattle, WA, 2556–2561.

[5] Chalon, M., Friedl, W., Reinecke, J.,Wimboeck, T., and Albu-Schaeffer, A. (2011). "Impedance control of a non-linearly coupled tendon driven thumb," in *Proceeding of 2011 IEEE/RSJ International Conference on Intelligent Robots and Systems*, San Francisco, CA, 4215–4221.

[6] Müller, U. K., and Van Leeuwen, J. L. (2006). Undulatory fish swimming: from muscles to flow. *Fish Fish.* 7, 84–103.

[7] Yuse, K., Guyomar, D., Audigier, D., Eddiai, A., Meddad, M., and Boughaleb, Y. (2013). Adaptive control of stiffness by electroactive polyurethane. *Sens. Actuators A Phys.* 189, 80–85.

[8] Luo, Q., and Tong, L. (2013). Adaptive pressure-controlled cellular structures for shape morphing I: design and analysis. *Smart Mater. Struct.* 22, 055014.

[9] Laffranchi, M., Tsagarakis, N. G., and Caldwell, D. G. (2013). Compact arm: a compliant manipulator with intrinsic variable physical damping. *Robotics* 8, 225–232.

[10] Erden, M. S., and Billard, A. (2015). Hand impedance measurements during interactive manual welding with a robot. *IEEE Trans. Robot.* 31, 168–179.

[11] Burgner-Kahrs, J., Rucker, D. C., and Choset, H. (2015). "Continuum robots for medical applications: a survey. *IEEE Trans. Robot.* 31, 1261–1280.

[12] Cianchetti, M., and Menciassi, A. (2017). "Soft Robots in Surgery," in *Soft Robotics: Trends, Applications and Challenges*: *Biosystems and Biorobotics*, Vol. 9, 1st Edn, eds C. Laschi, J. Rossiter, F. Iida, M. Cianchetti and L. Margheri (Cham: Springer International Publishing), 75–85.

[13] Suzumori, K., Iikura, S., and Tanaka, H. (1991). "Flexible microactuator for miniature robots," in *Proceedings of the IEEE Micro Electro Mechanical Systems*, Nara, 204–209.

[14] McMahan, W., Chitrakaran, V., Csencsits, M., Dawson, D., Walker, I. D., Jones, B., et al. (2006). "Field trials and testing of the OcotArm continuum manipulator," in *Proceedings of the 2006 IEEE International Conference on Robotics and Automation (ICRA)*, 2336–2341.

[15] Fraś, J., Czarnowski, J., Maciaś, M., Główka, J., Cianchetti, M., and Menciassi, A. (2015). "New STIFF-FLOP module construction idea for improved actuation and sensing," in *Proceedings of 2015 IEEE International Conference on Robotics and Automation (ICRA)*, Seattle, WA, 2901–2906.

[16] Cianchetti, M., Ranzani, T., Gerboni, G., De Falco, I., Laschi, C., and Menciassi, A. (2013). "STIFF-FLOP surgical manipulator: mechanical design and experimental characterization of the single module," in *Proceedings of the IEEE International Conference on Intelligent Robots and Systems (IROS)*, Tokyo, 3576–3581.

[17] Ranzani, T., Cianchetti, M., Gerboni, G., Falco, I. D., Petroni, G., and Menciassi, A. (2013). "A modular soft manipulator with variable stiffness," in *Proceedings of the 3rd Joint Workshop on New Technologies for Computer/Robot Assisted Surgery*, Verona.

[18] Suzumori, K., Maeda, T., Watanabe, H., and Hisada, T. (1997). "Fiberless flexible microactuator designed by finite-element method. *IEEE/ASME Trans. Mechatron.* 2, 281–286.

[19] Marchese, A. D., and Rus, D. (2016). Design, kinematics, and control of a soft spatial fluidic elastomer manipulator. *Int. J. Robot. Res.* 35, 840–869.

[20] Blanc, L., Delchambre, A., and Lambert, P. (2017). Flexible medical devices: review of controllable stiffness solutions. *Actuators* 6:23.

[21] Sareh, S., Jiang, A., Faragasso, A., Noh, Y., Nanayakkara, T., Dasgupta, P., et al. (2014). "Bio-inspired tactile sensor sleeve for surgical soft manipulators," in *Proceeding of 2014 IEEE International Conference on Robotics and Automation (ICRA)*, Hong Kong, 1454–1459.

[22] Manti, M., Cacucciolo, V., and Cianchetti, M. (2016). Stiffening in soft robotics: a review of the state of the art. *IEEE Robot. Automat. Magaz.* 23, 93–106.

[23] Wall, V., Deimel, R., and Brock, O. (2015). "Selective stiffening of soft actuators based on jamming," in *Proceeding of 2015 IEEE International Conference on Robotics and Automation (ICRA)*, Seattle, WA, 252–257.

[24] Cheng, N. G., Lobovsky, M. B., Keating, S. J., Setapen, A. M., Gero, K. I., Hosoi, A. E., et al. (2012). "Design and analysis of a robust, low-cost, highly articulated manipulator enabled by jamming of granular media," in *Proceedings—IEEE International Conference on Robotics and Automation*, Saint Paul, MN. 4328–4333.

[25] Jiang, A., Ataollahi, A., Althoefer, K., Dasgupta, P., and Nanayakkara, T. (2012). "A variable stiffness joint by granular jamming," in *Proceedings of the ASME Design Engineering Technical Conference, Parts A and B*, Vol. 4, Chicago, IL, 267–275.

[26] Hauser, S., Eckert, P., Tuleu, A., and Ijspeert, A. (2016). "Friction and damping of a compliant foot based on granular jamming for legged robots," in *Proceedings of the 2016 6th IEEE International Conference on Biomedical Robotics and Biomechatronics (BioRob)*, Singapore, 1160–1165.

[27] Stanley, A. A., and Okamura, A. M. (2016). "Deformable model-based methods for shape control of a haptic jamming surface. *Visual. Comput. Graph. IEEE Trans.* 23, 1029–1041.

[28] Santiago, J. L. C., Godage, I. S., Gonthina, P., and Walker, I. D. (2016). "Soft robots and kangaroo tails: modulating compliance in continuum structures through mechanical layer jamming. *Soft Robot.* 3, 54–63.

[29] Jiang, A., Aste, T., Dasgupta, P., Althoefer, K., and Nanayakkara, T. (2013). Granular jamming transitions for a robotic mechanism. *AIP Confer. Proc.* 1542, 385–388.

[30] Cheng, N., Amend, J., Farrell, T., Latour, D., Martinez, C., Johansson, J., A. et al. (2016). Prosthetic jamming terminal device: a case study of untethered soft robotics. *Soft Robot.* 3, 205–212.

[31] Sadati, S., Noh, Y., Naghibi, S. E., Kaspar, A., and Nanayakkara, T. (2015). "Stiffness control of soft robotic manipulator for minimally invasive surgery (MIS) using scale jamming," in *Proceedings of the International Conference on Intelligent Robotics and Applications (ICIRA), Lecture Notes in Computer Science*, (Portsmouth: Springer International Publishing), 141–151.

[32] Ataollahi, A., Karim, R., Fallah, A. S., Rhode, K., Razavi, R., L. Seneviratne, D., et al. (2016). Three-degree-of-freedom MR-compatible multisegment cardiac catheter steering mechanism. *IEEE Trans. Biomed. Eng.* 63, 2425–2435.

[33] Van Meerbeek, I. M., Mac Murray, B. C., Kim, J. W., Robinson, S. S., Zou, P. X., Silberstein, M. N., et al. (2016). Morphing metal and elastomer bicontinuous foams for reversible stiffness, shape memory, and self-healing soft machines. *Adv. Mater.* 28, 2801–2806.

[34] Shintake, J., Schubert, B., Rosset, S., Shea, H., and Floreano, D. (2015). "Variable Stiffness Actuator for Soft Robotics Using Dielectric Elastomer and Low-Melting-Point Alloy Soft state," in *Proceedings of the IEEE/RSJ International Conference on Intelligent Robots and Systems (IROS)*, Hamburg, 1097–1102.

[35] Tonazzini, A., Mintchev, S., Schubert, B., Mazzolai, B., Shintake, J., and Floreano, D. (2016). Variable stiffness fiber with self-healing capability. *Adv. Mater.* 28, 10142–10148.

[36] Jeong, S. H. (2016). *Liquids Matter in Compliant Microsystems*. Uppsala: Uppsala University.

[37] Janbaz, S., Hedayati, R., and Zadpoor, A. A. (2016). Programming the shape-shifting of flat soft matter: from self-rolling/self-twisting materials to self-folding origami. *Mater. Horiz.* 3, 536–547.

[38] Alambeigi, F., Seifabadi, R., and Armand, M. (2016). "A continuum manipulator with phase changing alloy," in *Proceedings of the IEEE International Conference on Robotics and Automation*, Stockholm, 758–764.

[39] Wang, W., Rodrigue, H., and Ahn, S.-H. (2016). "Deployable soft composite structures. *Sci. Rep.* 6, 20869–20869.

[40] Cheng, N. G., Gopinath, A., Wang, L., Iagnemma, K., and Hosoi, A. E., (2014). Thermally tunable, self-healing composites for soft robotic applications. *Macromol. Mater. Eng.* 299, 1279–1284.

[41] Yang, Y., and Chen, Y. (2016). "Novel design and 3d printing of variable stiffness robotic fingers based on shape memory polymer," in *Proceedings of the 2016 6th IEEE International Conference on Biomedical Robotics and Biomechatronics (BioRob)*, Singapore, 195–200.

[42] Yang, Y., Chen, Y., Li, Y., and Chen, M. Z. (2016). "3d printing of variable stiffness hyper – redundant robotic arm," in *Proceedings of the 2016 IEEE International Conference on Robotics and Automation (ICRA)*, Stockholm, 3871–3877.

[43] Lipton, J. I., and Lipson, H. (2016). 3d printing variable stiffness foams using viscous thread instability. *Sci. Rep.* 6:29996.

[44] Yang, Y., Chen, Y. H., Wei, Y., and Li, Y. (2016). Novel design and 3d printing of variable stiffness robotic grippers. *J. Mech. Robot.* 8, 134–143.

[45] Yuen, M. C., Bilodeau, R. A., and Kramer, R. K. (2016). Active variable stiffness fabric. *IEEE Robot. Autom. Lett.* 1, 708–715.

[46] Yao, Y., Zhou, T., Wang, J., Li, Z., Lu, H., Liu, Y., et al. (2016). 'Two way' shape memory composites based on electroactive polymer and thermoplastic membrane. *Comp. A Appl. Sci. Manuf.* 90, 502–509.

[47] Shiva, A., Stilli, A., Noh, Y., Faragasso, A., Falco, I. D., Gerboni, G., et al. (2016). Tendon-based stiffening for a pneumatically actuated soft manipulator. *IEEE Robot. Autom. Lett.* 1, 632–637.

[48] Afrisal, H., Sadati, S., and Nanayakkara, T. (2016). A bio-inspired electro-active velcro mechanism using shape memory alloy for wearable and stiffness controllable layers," in *Proceedings of the 2016 9th International Conference on Information and Automation for Sustainability (ICIAfS)* (Rome: IEEE), 1–6.

[49] Kretschmann, D. (2003). Nature materials: velcro mechanics in wood. *Nat. Mater.* 2, 775–776.

[50] Xu, X., Zhang, Q., Yu, Y., Chen, W., Hu, H., and Li, H. (2016). Naturally dried graphene aerogels with superelasticity and tunable poisson's ratio. *Adv. Mater.* 28, 9223–9230.

[51] Kwon, G. H., Park, J. Y., Kim, J. Y., Frisk, M. L., Beebe, D. J., and Lee, S. H. (2008). Biomimetic soft multifunctional miniature aquabots. *Small* 4, 2148–2153.

[52] Ebara, M., Kotsuchibashi, Y., Narain, R., Idota, N., Kim, Y.-J., Hoffman, J. M., et al. (2014). *Smart Biomaterials*. Berlin: Springer.

[53] Jones, C. D., and Steed, J. W. (2016). Gels with sense: supramolecular materials that respond to heat, light and sound. *Chem. Soc. Rev.* 45:c6cs00435k.

[54] Taylor, D. L., and In Het Panhuis, M. (2016). Self-healing hydrogels. *Adv. Mater.* 28, 9060–9093.

[55] Kuckling, D. (2009). Responsive hydrogel layers—From synthesis to applications. *Coll. Polym. Sci.* 287, 881–891.

[56] Osada, Y., and Gong, J.-P. (1998). Soft and wet materials: polymer gels. *Adv. Mater.* 10, 827–837.

[57] Palleau, E., Morales, D., Dickey, M. D., and Velev, O. D. (2013). Reversible patterning and actuation of hydrogels by electrically assisted ionoprinting. *Nat. Commun.* 4:2257.

[58] Park, K. (1988). Enzyme-digestible swelling hydrogels as platforms for long-term oral drug delivery: synthesis and characterization. *Biomaterials* 9, 435–441.

[59] Giammona, G., Pitarresi, G., Cavallaro, G., Buscemi, S., and Saiano, F. (1999). New biodegradable hydrogels based on a photocrosslinkable modified polyaspartamide: synthesis and characterization. *Biochim. Biophys. Acta* 1428, 29–38.

[60] Martens, P., and Anseth, K. S. (2000). Characterization of hydrogels formed from acrylate modified poly(vinyl alcohol) macromers. *Polymer* 41, 7715–7722.

PART II

Creation and Integration
of Multiple Sensing Modalities

6

Optical Force and Torque Sensor for Flexible Robotic Manipulators

**Yohan Noh[1], Sina Sareh[2], Emanuele Lindo Secco[3]
and Kaspar Althoefer[4]**

[1]Centre for Robotics Research, Department of Informatics, King's College London, London, United Kingdom
[2]Design Robotics, School of Design, Royal College of Art, London, United Kingdom
[3]Dept. of Mathematics and Computer Science Liverpool Hope University, Liverpool, United Kingdom
[4]Advanced Robotics at Queen Mary, Queen Mary University of London, London, United Kingdom

Abstract

Robot-assisted Minimally Invasive Surgery (RMIS) typically uses a master–slave surgical configuration allowing surgeons to carry out surgical tasks remotely. In typical RMIS scenarios, the surgeons use visual information provided by a three-dimensional (3D) camera to interact with patients' internal body organs via haptic interface devices. However, visual occlusion is one of the major drawbacks of the surgical approaches relying on visual information indicating the need for physical or virtual presence of the sense of touch during the operation. A multiaxis force sensor was developed to be integrated into the structure of surgical robots to enable continuous monitoring of external forces applied to the robot's body, thereby assisting the surgeon in undertaking precise control actions using more accurate sensory information. In this chapter, we report on the STIFF-FLOP approach in design and implementation of a three-axis force sensor based on fiber optics. The sensing system has a hollow geometry, is immune to electrical noises and low cost, and hence, is suitable for integration into a wide range of medical and industrial manipulation devices.

6.1 Introduction

Control of articulated surgical instruments and robots [1], e.g., the STIFF-FLOP, requires precise sensing of the robots' shape and end-effector position [2–4], as well as applied external forces [5–9]. In general, the 3D camera embedded in the surgical instrument can provide visual feedback during surgery [1], but it does not completely eliminate the need for haptic perception attained by feeling the touch. The sensation of the patient's organs provides valuable information to the surgeon such as the consistency and health of the tissue and organs. In addition, in the robot-assisted surgery, the sense of touch could assist the surgeon in controlling the amount of the exerted force on the delicate tissue, in order to prevent any damage to the tissue. In summary,

- Haptic feedback is important since it can enhance the patient's safety and prevent dangerous after-effects following the surgical procedure.
- Additional information retrieved from further sensor modalities may assist the surgeon when using robots for surgical procedures.

In the EU STIFF-FLOP project, it was essential to integrate force and torque sensors into the proposed robots in order to provide sensory feedback on the external forces. Here we presented the development of three-axis force and torque sensors based on a fiber optic Light Intensity Modulation (LIM) approach [5–10]. The sensing mechanism enables embedded measurement of force and torque values via low-power low-noise electrical and optical components encased into a flexible structure that can safely interact with human body.

The sensor makes use of optoelectronics and fiber optic technologies. It was designed, calibrated, tested, and fully integrated within the soft continuum STIFF-FLOP robot arm. The design of the sensor was optimized in order to work in a range of values of the applied forces, which are comparable with typical values in surgical scenarios, according to medical requirement and specifications provided by medical specialists involved in the project. Although we report only on the integration of the two sensor systems into the STIFF-FLOP arm, the design, geometry, and structure of the sensor allow integration into a wider range of robotic manipulation systems.

6.2 Materials and Methods

6.2.1 Sensor Design Rational

The proposed F/T sensing systems should satisfy a number of key technological and medical requirements to be applicable in medical and surgical
procedures. These include its ability in multiaxial measurement of the force
and torque values (F_{range}: ± 5.0 [N]; T_{range}: ± 3.5 N*cm, respectively), consistent operation with low hysteresis, satisfying manipulator's size restrictions
(maximum sensor's external diameter is 32 mm), and compatibility with
intraoperative Magnetic Resonance Imaging (MRI), and similar diagnosis
techniques which can be used during the surgical procedure. Considering the
above restrictions and the technical capacity required for a surgical robot, the
STIFF-FLOP F/T sensors were developed with the following specifications:

- The sensor should be embedded between two mutually tangent segments
  of the STIFF-FLOP manipulator.
- It should have a ring-like hollow structure to allow for actuation pipes
  and electrical wires to be passed through the arm.
- The device should be capable of simultaneous measurement of three
  components of external forces and moments, namely, the longitudinal force (F_z) and the two torque components (M_x, and M_y – see
  Figure 6.3).

Such measurement abilities are important for estimating the external forces
applied on the arm. Note that a two-segment STIFF-FLOP arm can undergo
two three-directional bending in each segment that can also be combined with
elongation of the arm.

6.2.2 Sensor Configurations

The three-axis force sensing can be conceived by adopting two classes of
light intensity-based approaches: optoelectronic and optical fiber-based techniques. These technologies allow compact design and modular integration
between successive segments of the STIFF-FLOP arm and reduce concerns
on possible interferences with intraoperative diagnosis techniques.

In the first approach, three optoelectronic sensors, model QRE1113
(Fairchild Semiconductor Corp., USA), were used in combination with three
reflectors, i.e., the mirrors (Figures 6.1 and 6.2a); in the second approach,
three pairs of optical fibers were employed, and again, in combination with
three reflecting surfaces (Figures 6.1 and 6.2b). Those sensing elements
are integrated into a flexible ring-like structure made from ABS plastic

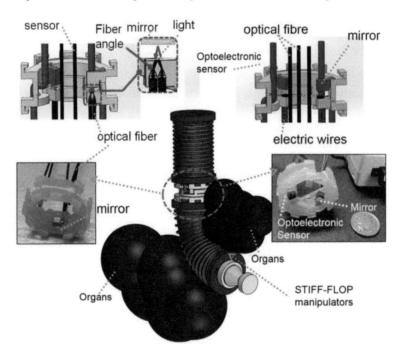

Figure 6.1   Force and torque sensors integrated into the STIFF-FLOP soft manipulator: fiber optic and optoelectronic technologies are shown on the left and right panels, respectively [5–9].

(copolymer of acrylonitrile, butadiene, and styrene) fabricated via a Projet^{TM} HD 3000 3D prototyping machine (3D Systems, Inc., USA) [5–9].

The optoelectronic type sensor consists of light-emitting diodes (LEDs) and phototransistors for emitting and receiving light, respectively (Figure 6.2a). The fiber optic type uses a pair of the two optical fibers connected to FS-N11MN fiber optic sensors (Keyence^{TM}, Japan), which is a similar LED-phototransistor arrangement integrated into a commercial product (Figure 6.2b).

In both cases, the three flexible cantilever beams were used as LIM mechanism to create the displacement of the element: they are equally distributed with an angular spacing of 120° (between any two) in the periphery of the sensor structure (Figures 6.1 and 6.3). The displacements of these cantilevers occur as soon as the sensor is stressed by a longitudinal load or lateral torques: the arrangement allows the measurements of the force component F_z and the two moment components, M_x and M_y.

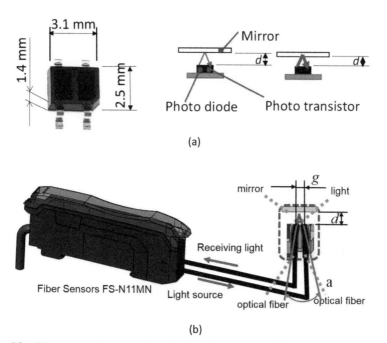

Figure 6.2 The two sensing approaches: the optoelectronic and fiber optic-based F/T sensors on the top (a) and bottom (b) panels, respectively [5–9].

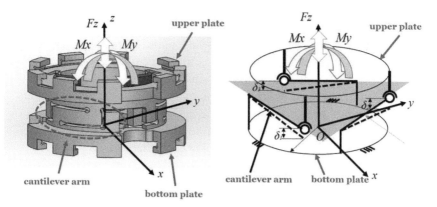

Figure 6.3 The F/T sensing principle based on three equally spaced flexible cantilever beams [5–9].

In the optoelectronic type sensor case, the light emitted from the LED is reflected by the mirror. The reflected light is transmitted to the photo-transistor which can convert light intensity to voltage values. In the same way,

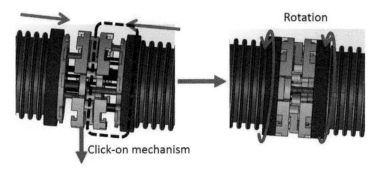

Figure 6.4 The integration into the STIFF-FLOP arm. The design allows insertion and removal of the sensor via a simple click-on mechanism [5–9].

in the fiber optic type case, emitting light and receiving light are transmitted via two fibers using KEYENCE fiber optic sensors (FS-N11MN) (Figure 6.2). The closer the distance between the reflector and the optoelectronic sensor or the distal end of the two fibers is, the higher is the amount of received light and its respective sensor output voltage. The experimental results show the reflected light intensity (converted to voltage values) changes linearly with respect to the distance changes between the optoelectronic sensor (or a pair of fibers) and the reflector [5–9].

It is worth noting that as some amounts of external force F_z and moments M_x and M_y are applied to the upper plate, the three associated cantilever beams are deflected. The three corresponding photo-interrupters, or three pairs of the optical fibers, measure the resultant cantilever beam deflections (δ_1, δ_2, and δ_3) between the upper and bottom plates of the sensor (Figure 6.3). From the three deflections, the external force F_z and moments M_x and M_y can be inferred by multiplying a calibration matrix [5–9].

In order to integrate the force sensor to the structure of the STIFF-FLOP manipulator, a locking and unlocking mechanism has also been integrated within the sensor design (Figure 6.4). Such a mechanism allows the integration of multiple sensors in between multiple modules of the robots.

6.3 Results and Discussion

In order to evaluate and test the performance of two types of developed sensors, external known forces and moments have been applied to the sensors using a custom characterization system [5, 8], in both optoelectronic and fiber optic configurations. The errors between the real values of the loads and the estimated ones using proposed sensors are shown in Tables 6.1 and 6.2.

Table 6.1 Sensor error property – optoeletronic technology

Force/Moment	Range	Maximum Error	Percentage Error [%]
F_z	±5.0 [N]	0.65 [N]	6.5
M_x	±3.5 [N.cm]	0.71 [N.cm]	10
M_y	±3.5 [N.cm]	0.45 [N.cm]	6.4

Table 6.2 Sensor error property – fiber optic technology

Force/Moment	Range	Maximum Error	Percentage Error [%]
F_z	± 3.0 [N]	0.32 [N]	10.7
M_x	±3.15 [N.cm]	0.38 [N.cm]	12.2
M_y	±3.15 [N.cm]	0.59 [N.cm]	18.2

The difference in the maximum error values for the two types of sensors is mainly due to different calibrations' complexities associated with the optoelectronic sensor (QRE1113) and the fiber force convertor (FS-N11MN). The optoelectronic sensor performance and its response can be tuned by changing d (the distance reported in Figure 6.2a), whereas the fiber optic sensor can be optimized by changing three different parameters g, a, and d (Figure 6.2b). Due to these inherent degrees of freedom of the mechanical design, each sensor prototype presents different values of these parameters and therefore requires its own specific calibration [5–9].

Both the optoelectronics and the fiber optic-based techniques have their advantages and drawbacks and, depending on the surgical application and required level of miniaturization within the surgical tool, they can be preferred one each other. One of the main advantages of the optoelectronic solution is its straightforward integration within the sensor structure and, consequently, into the robot. In case of the arm being damaged, the force/torque sensor can be removed and easily reinstalled again, and typically the calibration procedure does not need to be performed again. The main advantage of the fiber optic approach is the MRI compatibility and immunity from magnetic and electrical disturbances. In contrast, once the optical fiber is removed and reattached to the FS-N11MN convertor, typically, a recalibration of the sensor is required.

6.4 Conclusions

A novel force/torque sensor based on optoelectronic and fiber optic technologies for robotic and MIS has been presented. The sensor is particularly suited for integration into the articulated manipulation devices requiring pipes or tendons passing through the inner part of the sensor. The arm's proximal and distal segments can be equally connected to the base electrically and pneumatically, thanks to the sensor's hollow structure.

The results of experiments show maximum errors of around 18% and 6% in estimating external forces using fiber optics and optoelectronic configurations, respectively. Due to the variability of the geometry of the sensor prototype, the fiber optic configuration presents a quite large percentage error, which may be strongly reduced by standardizing the position and orientation of the fibers within the sensor framework. The proposed system can help in enhancing the haptic feedback in robot-assisted surgical procedures and palpation devices. Improvements on the sensor manufacturing process, material, and structure should be investigated to reduce the error and hysteresis, and enhance the sensor linearity.

Since the STIFF-FLOP project has ended, the team has been developing similar technologies for tactile and force/torque sensing for applications such as palpation tasks, automatic localization of tumors, flexible manipulators, and robotic hand fingertips for dexterous manipulation and grasping of objects [11–16].

References

[1] Simaan, N., Xu, K., Kapoor, A., Wei, W., Kazanzides, P., Flint, P., et al. (2009). A system for minimally invasive surgery in the throat and upper airways. *Int. J. Rob. Res.* 28, 1134–1153.

[2] Sareh, S., Noh, Y., Ranzani, T., Liu, H., and Althoefer, K. (2015). "A 7.5mm Steiner chain fiber-optic system for multi-segment flex sensing," in *Proceedings of the IEEE/RSJ International Conference on Intelligent Robots and Systems (IROS)* (Hamburg: IEEE), 2336–2341.

[3] Sareh, S., Noh, Y., Li, M., Ranzani, T., Liu, H., and Althoefer, K. (2015). Macro-bend optical sensing for pose measurement in soft robot arms, Smart Mater. *Structure* 24:125024.

[4] Wurdemann, H., Sareh, S., Shafti, A., Noh, Y., Faragasso, F., Liu, H., et al. (2015). "Embedded electro-conductive yarn for shape sensing of soft robotic manipulators," in *Proceedings of the Engineering in Medicine and Biology Conference (EMBC)* (Milan: IEEE).

[5] Noh, Y., Secco, E. L., Sareh, S., Wurdemann, H., Faragasso, A., Althoefer, K., et al. (2014). A continuum body force sensor designed for flexible surgical robotics devices. *Conf. Proc. IEEE Eng. Med. Biol. Soc.* 2014, 3711–3714.

[6] Noh, Y., Sareh, S., Ranzani, T., Faragasso, A., Liu, H., Althoefer, K., et al. (2014). "A three-axial body force sensor for flexible manipulators," in *Proceedings of the IEEE International Conference on Robotics and Automation (ICRA)* (Hong Kong: IEEE), 6388–6393.

[7] Noh, Y., Shiva, A., Hamid, E., Liu, H., Althoefer, K., and Rhode, K. (2016). "Light intensity based optical force/torque sensor for robotic manipulators," in *Proceedings of the 6th Joint Workshop on New Technologies for Computer/Robot Assisted Surgery* (Tempe: CRAS).

[8] Noh, Y., Sareh, S., Wurdemann, H., Liu, H., Housden, J., Althoefer, K., et al. (2015). Three-axis fiber-optic body force sensor for flexible manipulators. *IEEE Sens. J.* 16, 1641–1651.

[9] Noh, Y., Rhode, K., Bimbo, J., Sareh, S., Wurdemann, H., Fraś, J., et al. (2016). Multi-axis force/torque sensor based on simply-supported beam and optoelectronics. *Sensors* 16:1936.

[10] Sareh, S., Jiang, A., Faragasso, A., Noh, Y., Nanayakkara, T., Dasgupta, P., et al. (2014). "Bio-inspired tactile sensor sleeve for soft surgical manipulators," in *Proceedings of the IEEE International Conference on Robotics and Automation (ICRA)* (Hong Kong: IEEE).

[11] Noh, Y., Rhode, K., Luo, S., and Lam, Y.-T. (2017). "Modular tactile sensing array for localising a tumor for palpation instruments," in *Proceedings of the 7th Joint Workshop on New Technologies for Computer/Robot Assisted Surgery (CRAS)* (Tempe, AZ: CRAS), 6388–6393.

[12] Noh, Y., Rhode, K., Jan, F., and Gawenda, P. (2017). "Contact force sensor for flexible manipulators for MIS (minimally invasive surgery)," in *Proceedings of the 7th Joint Workshop on New Technologies for Computer/Robot Assisted Surgery (CRAS)* (Tempe, AZ: CRAS).

[13] Noh, Y., Rhode, K., Luo, S., and Lam, Y.-T. (2017). Human finger inspired grasping structure using tactile sensing array with single type optoelectronic sensor. *IEEE Sens.* 267, 18–24.

[14] Konstantinova, J., Cotugno, G., Stilli, A., Noh, Y., and Althoefer, K. (2017). Object classification using hybrid fiber optical force/proximity sensor. *IEEE Sens.* 2017:e0171706.

[15] Maereg, A. T., Secco, E. L., Fikire, T., Reid, D., and Nagar, A. (20174). A low-cost, wearable opto-inertial 6 DOF hand pose tracking system for VR. *Wearable Technol.* 5:49.

[16] Maereg, A. T., Nagar, A. K., Rcid, D., and Secco, E. L. (2017). Wearable vibrotactile haptic device for stiffness discrimination during virtual interactions. *Front. Robot. AI Biomed. Robot.* 4:42.

7

Pose Sensor for STIFF-FLOP Manipulator

**Sina Sareh[1], Yohan Noh[2], Tommaso Ranzani[3], Min Li[4]
and Kaspar Althoefer[5]**

[1]Design Robotics, School of Design, Royal College of Art, London,
United Kingdom
[2]Centre for Robotics Research, Department of Informatics, King's College
London, London, United Kingdom
[3] Department of Mechanical Engineering, Boston University, Boston, MA,
United States
[4]Institute of Intelligent Measurement and Instrument, School of Mechanical
Engineering, Xi'an Jiaotong University, Xian Shi, China
[5]Advanced Robotics at Queen Mary (ARQ), Faculty of Science and
Engineering, Queen Mary University, London, United Kingdom

Abstract

The STIFF-FLOP robotic arm is a cylindrical structure made from soft
silicone rubber materials encasing pneumatic actuation chambers. Its material
properties and structure allows for the shape, and therefore the pose of
the arm, to be dictated by the actuation system as well as by the surfaces
with which it is interacting. Although this softness and flexibility makes it
inherently safe for many medical and industrial applications, such as keyhole
surgery, it comes at the expense of complicating sensing and position control.
This chapter presents the main challenges for the development of a pose
sensor for soft robotic arms and the STIFF-FLOP approach to tackle them.

7.1 Introduction

Construction of robots using soft materials and components [1–9] promises
great potential particularly from the point of view of safe human–robot
interaction [10, 11]. However, it is faced with intriguing engineering
challenges with respect to configuration and position control [1, 12].

Conventional methods for calculating kinematics and dynamics of robots assume that they are made out of rigid material, their body may only bend where there are joints [13–18], and in the case of a collision with other hard structures and robots, the methods of rigid-body physics are applied [19]. However, the methods are not directly applicable to robots composed from soft and deformable materials, implying the need for bridging the gap by the development of the respective theoretical and experimental methods enabling precise sensing and control in soft robots.

Technologies for hyper-redundant [13–17] and soft [2–8, 20–25] robotic manipulation constitute an important category in soft robotics research with various industrial and medical applications, such as articulated robotic tools for operation in confined spaces, e.g., STIFF-FLOP [3, 19–25]. The tools are usually made from mutually-tangent curved segments enabling high degrees of robotic articulation in hard-to-access, unstructured, and cluttered workspaces [3, 26]. However, control of these robots requires precise sensing of the robot's pose—information on the position and orientation of the robot end-effector—and shape—information on the robot's articulated body form.

The complex shape generated by these robots have been mainly tracked through incorporation of vision systems [27] and electromagnetic tracking [28]; these visual techniques are often restricted with visual occlusion and electromagnetic tracking; they are subject to magnetic field distortions and have limitations with regard to the mobility of the magnetic field generation system. However, the required information on the complex shape and pose of the robot can be obtained through an appropriate multi-segment flex sensing method [29].

A number of sensing mechanisms for measuring the flexion have been proposed in the literature. Prominent examples include: off-the-shelf resistive flex sensors based on conductive ink, e.g., *FLXT* (Flexpoint Sensor Systems, Inc., United States); flexible sensors based on specific types of smart materials, e.g., *Ionic Polymer Metal Composite (IPMC)* [30]; soft sensors based on the micro-channel of conductive liquid (Eutectic Gallium Indium, *eGaIn*) [31]; and sensing systems based on fiber optics. Resistive sensors based on conductive inks and IPMCs are bipolar devices and are not usually suitable for three-dimensional fabrication. The sensing systems based on eGaIn are attractive for integration in soft structures and robots; however, there is no data on biocompatibility of this material according to the datasheet published. Sensors based on fiber optics function by measuring the change in optical characteristics of the light [4, 19, 26, 32–37]. From the electrical point of view, optical fibers are immune to magnetic fields and electrical interference and hence, they are distinguished candidates for many industrial and medical applications. From the mechanical point of view, plastic optical fibers are very attractive for integration into soft structures due to their ability to follow the elastic deformation of the robot bodies in which they are embedded. Optical sensors based on Fiber Bragg Grating are costly and sensitive to temperature and strain [34, 38].

STIFF-FLOP employed light intensity modulation to produce a low-cost multi-segment optical curvature sensor amenable to being integrated into for flexible, soft and extensible robotic arms [26, 29, 32]. In the following, the design challenges and STIFF-FLOP approaches in implementation and testing of the pose-sensing system are discussed; starting with sensing solutions for a one-segment STIFF-FLOP arm (also referred to as STIFF-FLOP module) as well as generalization of the method for use in a multi-segment manipulator.

7.2 Design of the Pose-sensing System

7.2.1 Pose-sensing in a One Segment STIFF-FLOP Arm

The pose sensor of each segment of the arm consists of three optical fibers sliding inside flexible housings in the periphery of the arm, as illustrated in Figures 7.1a and b, in parallel with actuation chambers. When the arm bends, the optical fibers' length portion inside the flexible arm (s_1, s_2, and s_3) will change according to the pose (amount of flexure and the tip orientation) of the arm; it causes a change in the position of the light-emitting optical fibers and, consequently, the intensity of the light received by light detectors, e.g., FS-N11MN (Keyence^{TM}, Japan). The received light is then converted into voltage $v = [v_1 \; v_2 \; v_3]$ and related to the corresponding distance vector $s = f(v) = [s_1 \; s_2 \; s_3]$, to acquire configuration parameters of the arm segment for each specific pose. The configuration parameters are as follows: S is the length of the central axis of each segment, θ is the bending angle, and φ is the orientation angle, and can be expressed as,

$$S = \frac{1}{3}\sum_{i=1}^{3}(S_i) \tag{7.1}$$

$$\theta = \frac{S - s_1}{d.\cos(\frac{\pi}{2} - \varphi)} \tag{7.2}$$

$$\varphi = \tan^{-1}\left(\frac{\sqrt{3}\,(s_2 + s_3 - 2s_1)}{3\,(s_2 - s_3)}\right) \tag{7.3}$$

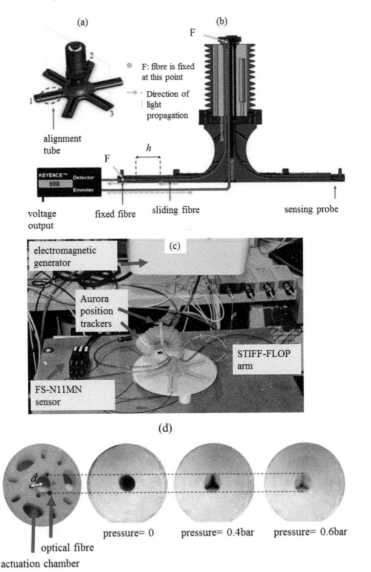

Figure 7.1   (a) The overall structure of the STIFF-FLOP module with integrated bending sensor, (b) the experimental configuration for measuring the flexure angle; four Aurora electromagnetic trackers are integrated at the base, tip and on the body of the arm, (c) the cross section of the module indicating the position of pneumatic actuators and optical fiber, and (d) the top view of the module indicating the negative impact of actuation chambers on the sensing system; increasing the pressure inside the actuation chambers results in a ballooning effect toward the internal hollow structure which changes the radial location of optical fibers used for pose-sensing.

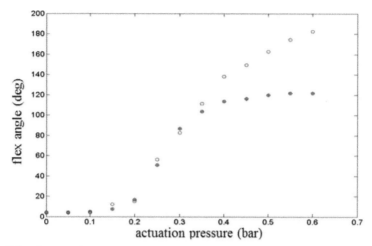

Figure 7.2   The experimental curvature sensing results for the experimental configuration described in Figure 7.1.

Variable d in Equation (7.2) describes the distance between the central axis of the arm segment and the parallel optical fibers. Being integrated into a soft structure, this distance between components of the sensing system may change during the operation of the arm by the internal actuation system or via externally applied forces and directly affect its resolution. The experimental setup for measuring the flexure angle is shown in Figure 7.1c where an Aurora electromagnetic tracking system is used for benchmarking. The experimental results (Figure 7.2) show an error of more than 20° when pressuring a single channel up to 0.4 bar with an increasing trend for larger amounts of input air pressure which is due to the movement of optical fibers by the parallel actuation chambers, as shown in Figure 7.1d.

To overcome the aforementioned problem, the STIFF-FLOP pose-sensing system considered the following: (1) in order to increase the robustness of the sensors, no optical fibers should be sent inside the soft arm and (2) an internal support structure that can maintain the radial location of the sensing components inside the soft structure of the arm should be added to protect the sensing system against the unwanted relative motion dictated by the actuation system.

Hence, a flexible internal structure responsible for maintaining the radial distance between passive cables and the center of the flexible arm during manipulations at the distal side was created along with a distance modulation array which couples the motion of passive cables with light-emitting optical

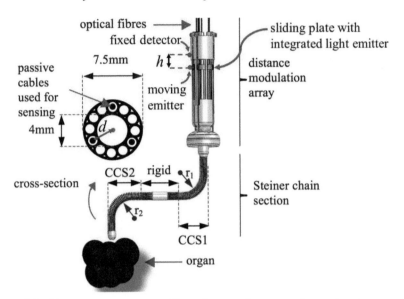

Figure 7.3   The two-segment arm with an integrated pose-sensing system in a surgical scenario interacting with an organ. The flexible Steiner chain section also provides an empty central channel for end-effector tools (CCS1 and CCS2 indicate constant curvature segments with radii of r_1 and r_2, respectively).

fibers using a low-friction sliding mechanism at the proximal side. The light-emitting optical fiber is paired and aligned with optical detectors fixed at the base of the arm, as illustrated in Figure 7.3.

The pose-sensing principle employs multiple passive cables passed through 1.2 mm (outer diameter) spring channels integrated along the length of the arm (50 mm in length). These channels are located at the same distance d from the central axis of the arm but using different equally spaced angular positions. The cable channels are continued outside the arm using the 3D printed part of the sensor where they are converted to sliding rails. A very low-friction sliding mechanism was created employing two steel needles with thickness of 0.89 mm located parallel to and 4 mm away from each other. A specialized U-shaped mechano-optical coupler[1] was designed and fabricated to be able to smoothly slide around needles and carry the light-transmitting optical fibers inside the sensor base. The sliding plate is linked with a 2 mm outer diameter extension spring which is fixed at one side and

---

[1]The interface between passive cables of the Steiner chain section and optical fibers of the distance modulation array.

enables retraction and position recovery after being pulled by cables. This arrangement allows the pose sensor to exploit a retractable sliding mechanism for modulating the distance between emitter and detector optical fibers.

7.2.2 The Flexible Steiner Chain Section

In geometry, a Steiner chain is a set of mutually tangent n circles, all of which are tangent to two given non-intersecting circles, as illustrated in Figure 7.4a. The Steiner chain mathematics can explain the design of a steerable endoscope in [39], using low-cost commercial springs; the springs are tangentially combined in parallel to tightly accommodate driving tendons and prevent their radial displacement. Figure 7.4a also describes the cross-section of the endoscopic mechanism; note that the structure can also house passive cables to code the shape of the endoscope or similar manipulation systems. In the following, we will report on the Steiner chain implementation of STIFF-FLOP's pose-sensing system and the details of the new design.

Referring to Figure 7.4b and Equations (7.1) to (7.3), the amount of change in values of s_1, s_2, and s_3 due to a bending is in a direct relationship with d. Therefore, d is directly affecting the resolution of the sensor system: By substituting $\varphi_1 = \frac{\pi}{2}$ and a very small amount of bending $\theta_1 = \frac{\pi}{180}$ in Equations (7.3) and then (7.2)

$$- 2s_{1,1} + s_{2,1} + s_{3,1} = 0.0525d \qquad (7.4)$$

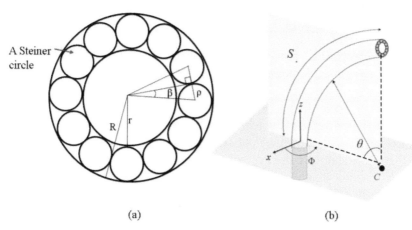

(a)   (b)

Figure 7.4 (a) The Steiner chain cross-section of the arm and (b) the flexed configuration of the arm and associated parameters in 3D [38].

where $s_{i,1}$ denotes the primary length of the cable i which is inside the arm segment. Also from Equation (7.3), $s_{2,1} = s_{3,1}$, therefore, it yields

$$s_{2,1} - s_{1,1} = 0.0262d \tag{7.5}$$

If substituting $\varphi_2 = \frac{\pi}{2}$ and $\theta_2 = \frac{\pi}{2}$ in the same set of equations

$$s_{2,2} - s_{1,2} = \frac{3\pi}{4}d, \tag{7.6}$$

where $s_{i,2}$ explains the secondary length of the cable i which is inside the arm segment. Therefore the total change in the length of the cable s_1 can be obtained by subtracting Equations (7.5) and (7.6)

$$\Delta s_1 = s_{2,2} - s_{1,2} - s_{2,1} + s_{1,1} \tag{7.7}$$

The arm segments are made from extension springs and therefore are incompressible. Assuming $s_{2,2} \approx s_{2,1}$, the change in the cable's length inside the arm can be calculated through

$$\Delta s_1 \approx \frac{3\pi}{4}d \tag{7.8}$$

The KEYENCE FS-N11MN fiber optic light-to-voltage convertor used in this study can effectively measure a maximum fiber length change of 20 mm. In order to measure a 90° bending deformation, the maximum value of d (d_{max}) must not exceed 8.4 mm. It is clear that, this value should be reduced to 4.2 mm for measuring a maximum of 180° bending which can be regarded as two successive 90° bending as targeted in this study. Since the maximum combined deformation of the two segments should be also measureable within the 20 mm range, the maximum value of the d parameter, d_{max}, needs to be 2.1 mm.

In order to preserve the maximum resolution of the sensing system, we choose the maximum value for d, which is 2.1 mm. This needs incorporating of an inner spring with a diameter of slightly less than 4 mm. Therefore a LEM050AB 05 S stainless steel extension spring (Lee Spring Ltd., United States) with an outer diameter of 3.505 mm, a wire diameter of 0.508, and a stiffness rate of 0.04 N/mm was used as the central spine. To implement Steiner springs which correspond to Steiner circles in the cross-section view, see Figure 7.4a, custom springs with an outer diameter of 1.2 mm and a wire diameter of 0.25 mm were used. Steel passive cables with a diameter of 0.27 mm (Carl Stahl Ltd., Germany) were radially fixed at approximately

d = 2.1mm away from the center. The central angle $\beta = \sin^{-1}\left(\frac{\rho}{r+\rho}\right)$ is approximately 15°, the number of Steiner springs is $n = \frac{180}{\beta} = 12$, and the inner diameter of the outer spring is approximately $R = r + 2\rho = 5.91$ mm. Therefore, LEM070CB 05 S (Lee Spring Ltd., United States) with an outer diameter of 7.49 mm and a wire diameter of 0.711 mm was selected.

7.2.3 Design of a Low-friction Retractable Distance Modulation Array

When the arm bends, the portion of length of each passive cable which is inside the arm will change. As it returns to its original straight configuration, the cables' length portion inside the arm are also required to return to their original state. Several reasons including friction, and hysteresis in the mechanical structure and material properties prevent meeting this essential condition and introduce malfunction into the pose-sensing system. In order to make sure the mutual distance between the emitting and detecting fiber optic pairs is recovered, a spring returning mechanism is a straightforward solution. In addition, this mechanism couples the motion of passive cables that are passed through the length of the flexible arm with optical fibers for light intensity measurement using KEYENCE convertors.

7.2.3.1 Loopback design of the optical system

The *commercial off-the-shelf* stretch (length) sensors are usually fabricated to be free from electronics at one end. Examples of such implementation include stretch sensors from StretchSense Ltd, New Zealand, and PolyPower® Stretch Sensors. This free end is usually coupled with the moving end of the actuator to measure the length change. In our work, in order to allow all the electronics to be at one end of the sensor, we used a U-shape arrangement of optical fibers to produce a loopback configuration. This enables placing the emitter and detector next to each other, as illustrated in Figure 7.5b.

7.2.3.2 Steel spring-needle double slider

Modulating the mutual distance between any pair of optical fibers required a low-friction sliding mechanism to be designed and implemented. We manufactured a highly smooth double slider, which uses two steel needles (44 mm length × 0.86 mm diameter, John James Needles, Worcestershire, England). The needles were surrounded by two pieces of miniature steel springs (1.4 mm outer diameter, 0.2 mm wire diameter) installed in parallel

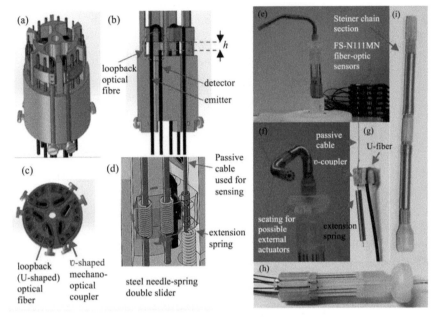

Figure 7.5   The CAD designs of the multi-segment flex sensor exhibiting integrated technologies: (a) the side view showing the elastic recovery mechanism of the slider, (b) the low-friction steel spring-double needle sliding mechanism to maintain the direction of the motion, (c) the U-shaped fiber optic arrangement, (d) close-up view of the sliding mechanism. The prototype of the arm and sensing system: (e) the finished configuration of the sensorized manipulation system, (f) close-up view of the top part, (g) the U-shaped mechano-optical coupler, (h) the fully assembled structure of the distance modulation array, and (i) the structure of the Steiner chain section.

and 4 mm away from each other into the plastic sensor base. Each steel spring that was able to smoothly slide around the surrounded needle, was embedded into a U-shaped plastic mechano-optical coupler. Each coupler linked a passive cable with its associated fiber optic pair. This mechanism is shown in Figure 7.5d. Note that Figure 7.5c shows the cross-section of the distance modulation array, highlighting its internal structure.

7.3 Fabrication and Assembly of the Pose-sensing System

The sensorized arm in its finished configuration is shown in Figures 7.5e and f. The structure of the mechano-optical coupler is shown in Figure 7.5g. It comprises a U-shaped 3D printed part with a U-shaped housing for an optical fiber, and a connection module which connects a 2 mm outer diameter extension spring (Lee Spring Ltd., United States), with a 0.27 mm thick steel wire rope (Carl Stahl, Germany), used as the passive cable. The 2 mm extension springs are in charge of cables position recovery and therefore are referred to as recovery springs. In order to increase the pulling force, an initial stretch of 5 mm for the recovery springs was considered. Finally, Figures 7.5h and i describe the full assembled structure of the main Steiner chain, and distance modulation array parts of the flexible arm. All plastic parts were manufactured using 3D printing (Projet HD-3000 Plus 3D Systems).

7.4 Sensor Calibration and Benchmarking

A set of experiments were performed to validate the design and implementation of the two-segment pose sensor in which either one or both of the segments were actuated at a time. Two high-definition (HD) cameras were placed at the top and side of the arm to record ground truth flex information. The middle and tip of the arm were attached to fixed points on the wall, using steel wires to generate stable shape patterns, as our experimental prototype was not yet equipped with motors or any other actuation system. Subsequently we have recorded the light intensity (and respective voltage values) from KEYENCE optical convertors and HD cameras.

To convert voltage values to corresponding values of distance between optical fiber tips h, the calibration relationships for all six fiber optic channels were extracted. The averaged calibration data, presented in Figure 7.6, was splined using MATLAB (MathWorks, Inc., Natick, MA, United States) software to form the calibration curve (over five trials).

After the calibration of sensors, the arm was forced into various 3D shapes, as shown in Figure 7.7, and the sensor voltage signals were recorded for analysis. In order to use the constant curvature bending model, the acquired voltage signals were fed into the splined calibration curve of Figure 7.6 to back-calculate the tip-to-tip fiber optic distances. Then, these distances were substituted into Equations (7.2) and (7.3) to calculate the flexion of the arm. Figure 7.8b shows the experimental results compared with their respective ground truth information extracted from HD camera images using a custom MATLAB code.

Figure 7.8a, $s_{i,j}$ represents the tip-to-tip distance between the optical fiber pair, where $i = \{1,2,3\}$ is the pair number and $j = \{1,2,3,4,5\}$ is the trial number. The trend in $\alpha = s_{i,j}/V_{i,j}$ implies that the arm was bent approximately symmetric with respect to cables b_1 and b_3, where these

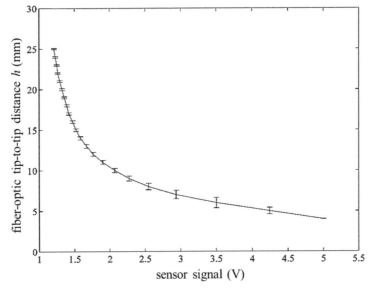

Figure 7.6 The averaged calibration curve and error bars.

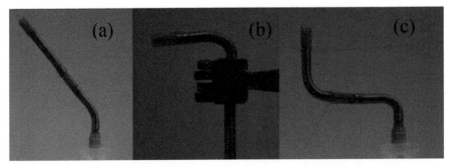

Figure 7.7 Examples of key experimental configurations: (a) independent activation of base segment, (b) independent activation of the tip segment (note that the intersegment link is clamed to produce a stationary base), and (c) simultaneous activation of two segments.

two cables are virtually stretched (note that only the length portion of the cables inside the arm can change physically). The cable b_2 was only slightly compressed, with respect to the length change in the other two cables which confirms the design assumption that led to Equation (7.6).

Figure 7.8b shows the experimental results where segments of the arm were bent individually, which implies a maximum tracking error of around $7°$ in the tip segment when the arm was bent with ground truth value of $40°$.

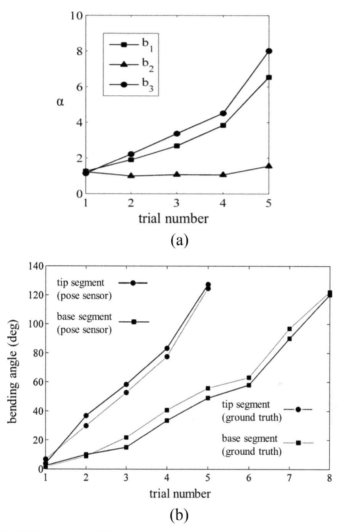

Figure 7.8 (a) The $\alpha = s_{i,j}/V_{i,j}$ values for the base segment experiment, b_n is a cable number, and (b) the experimental results of flex sensing in individual segments.

This error was decreased to 5° as the arm reached a ground truth flexion of 63° and around 2° for ground truth value of 122°. A similar behavior can be seen in the flex data of the base segment, also presented in Figure 7.7b. The finer function of the sensor for large amounts of flexion can be relevant for better positioning of cables inside the spring channels. The channels have an internal diameter of around 0.8 mm and house two 0.27 mm-thick cables along the length of the base segment, and only one of them along the length of the tip segment, which gives some room for the cables to play radially, if they are not pulled tightly using extension springs. The experimental results imply that the error can be reduced if this initial stretch is increased from the 5mm (the original amount of initial stretch in our design).

7.5 Calculation of the Bending Curvature in a Two-segment Arm Based on Collocated Cables

To simplify the multi-segment flexion sensing, this work used a method referred to as "collocated cables." This sensing arrangement uses only 3 Steiner channels out of the 12 for sensing. There are two passive cables sliding inside each of these three channels; one fixed between two segments and the other one at the tip of the arm. This arrangement enables measuring the flex angle in multi-segment arms in a modular way and with minimal amendment in the cables' length computation. When the arm undergoes a complex two-segment movement, the change in the length of cables passing through the whole length of the arm to the tip (t-type cables t_1, t_2, and t_3), is only partially because of the flexion of the tip segment. However, the change in the length of the fibers fixed at the tip of the base segment (b-type cables b_1, b_2, and b_3) is purely due to the flexion in this segment. In our sensing arrangement, each t-type cable is accompanied by a b-type cable, as shown in Figure 7.9. This method allows calculating the share of each segment from the total length change.

To intuitively evaluate this method, the segments of the arm were forced into a complex S-shape configuration, as shown in Figure 7.6c. Using camera ground truth information, we have calculated the flexion angle in two segments as $\theta_{\text{Ground},1} = 91.2°$, $\theta_{\text{Ground},2} = 95.1°$.

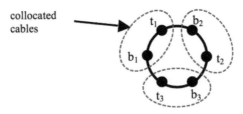

Figure 7.9   The cross-section of the distance modulation array, showing the collocated arrangement of passive cables used for sensing.

Table 7.1 Calculation of the pure length change in each segment based on collocated cables approach

Voltage (V)	V_{b1}	V_{b2}	V_{b3}	V_{t1}	V_{t2}	V_{t3}
	3.99	2.92	1.93	1.89	2.27	1.93
Length (mm)	s_{b1}	s_{b2}	s_{b3}	s_{t1}	s_{t2}	s_{t3}
	5.34	7.04	10.87	11.11	9.01	10.87
Length (mm)	s_{p1}		s_{p2}		s_{p3}	
	5.77		1.96		0	

Table 7.1 summarizes the calculation of length change in each segment based on the collocated cables' lengths; V_{bi} values, i = {1,2,3}, represent FS-N11MN voltage readings associated with cables that are fixed between two segments; V_{ti} are readings associated with cables passed through the whole length of the arm and fixed at the tip. Mapping into the splined voltage-distance relationship (Figure 7.6), the corresponding fibers' mutual distances s_{bi} and s_{ti} were calculated. Whilst values of s_{bi} represent the pure length change in the base segment, the pure length change of the tip segment can be computed through $s_{pi} = s_{ti} - s_{bi}$.

By substituting pure distance values into Equations (7.1) to (7.3), the flexion angles are computed as $\theta_{Sensor,1} = 89.2°$ and $\theta_{Sensor,2} = 92.4°$, implying an error of less than 3° in each segment.

7.6 Conclusion

In this chapter, we have presented the design and implementation of a pose-sensing system for soft robot arms. Starting from theoretical design, the radial location of the passive cables used for sensing along the periphery of the arm were optimized. This optimization work presents a trade-off between maximum compactness of the sensing system and using the full resolution of the optical measurement system. In the next step, we have presented a Steiner chain design for the flexible part of the sensor system. Three (out of twelve) Steiner chain channels were used for pose-sensing in the two-segment flexible arm. Subsequently, a low-friction fiber-optic distance modulation array based on a new spring-needle double slider we designed and implemented to precisely measure the change in the length of cables embedded in the periphery of the arm. The sensing system also features a loopback optical design to keep all electronics away from of the sensing site.

The sensing system is experimentally validated resulting in a maximum error of 6°, with respect to the camera ground truth information in measuring the flexion angle in individual segments. From the experimental results, it can be implied that the sensor's error can be reduced by increasing the initial stretch length of recovery springs, to make a tighter cabling system in low flexion. We have also demonstrated and discussed multi-segment flex sensing using collocation of passive cables in mutually-tangent (successive) segments. This sensing system can be regarded as complementary to the two-segment soft actuation system presented in [20].

Acknowledgment

The work described in this paper is funded by the Seventh Framework Programme of the European Commission under grant agreement 287728 in the framework of EU project STIFF-FLOP.

References

[1] Rus, D., and Tolley, M. T. (2015). Design, fabrication and control of soft robots. *Nature* 521, 467–475.

[2] Trivedi, D., Rahn, C. D., Kier, W. M., and Walker, I. D. (2008). Soft robotics: biological inspiration, state of the art, and future research. *Appl. Bionics Biomech.* 5, 99–117.

[3] Cianchetti, M., Ranzani, T., Gerboni, G., Nanayakkara, T., Althoefer, K., Dasgupta, P., et al. (2014). Soft robotics technologies to address shortcomings in today's minimally invasive surgery: the STIFF-FLOP approach. *Soft Robot.* 1, 122–131.

[4] Sareh, S., Althoefer, K., Li, M., Noh, Y., Tramacere, F., Sareh, P., et al. (2017). Anchoring like octopus: biologically inspired soft artificial sucker. *J. R. Soc. Interface* 14:20170395.

[5] Kim, S., Laschi, C., and Trimmer, B. (2013). Soft robotics: a bioinspired evolution in robotics. *Trends Biotechnol.* 31, 287–294.

[6] Stilli, A., Wurdemann, H. A., and Althoefer, K. (2014). "Shrinkable stiffness-controllable soft manipulator based on a bio-inspired antagonistic actuation principle," in *Proceedings of the IEEE/RSJ International Conference on Intelligent Robots and Systems*, Madrid.

[7] Sareh, S., Conn, A. T., and Rossiter, J. M. (2010). "Optimization of bio-inspired multi-segment IPMC cilia," in *Proceedings of the Electroactive Polymer Actuators and Devices (EAPAD XII)*, Vol. 7642, (Bellingham, WN: SPIE).

[8] Rossiter, J., and Sareh, S. (2014). "Kirigami design and fabrication for biomimetic robotics," in *Proceedings of the SPIE 9055, Bioinspiration, Biomimetics, and Bioreplication, 90550G* (Bellingham, WN: SPIE).

[9] Kim, J., Alspach, A., and Yamane, K. (2015). "3d printed soft skin for safe human-robot interaction," in *Proceedings of the Intelligent Robots and Systems (IROS), 2015 IEEE/RSJ International Conference on IEEE*, Madrid, 2419–2425.

[10] Schiavi, R., Grioli, G., Sen, S., and Bicchi, A. (2008). "Vsa-ii: a novel prototype of variable stiffness actuator for safe and performing robots interacting with humans," in *Proceedings of the IEEE International Conference on Robotics and Automation* (Rome: IEEE), 2171–2176.

[11] Marchese, A. D., Komorowski, K., Onal, C. D., and Rus, D. (2014). "Design and control of a soft and continuously deformable 2D robotic manipulation system," in *Proceedings of IEEE International Conference on Robotics and Automation,* Singapore.

[12] Yamada, H., Chigisaki, S., Mori, M., Takita, K., Ogami, K., and Hirose, S. (2005). "Development of amphibious snake-like robot ACM-R5," in *Proceedings of the 36th International Symposium Robotics*, Tokyo, 133.

[13] Lipkin, K., Brown, I., Peck, A., Choset, H., Rembisz, J., Gianfortoni, P., et al. (2007). "Differentiable and piecewise differentiable gaits for snake robots," in *Proceedings of the IEEE International Conference on Intelligent Robots and Systems*, San Diego, CA, 1864–1869.

[14] Shang, J., Noonan, D., Payne, C., Clark, J., Sodergren, M. H., Darzi, A., et al. (2011). "An articulated universal joint based flexible access robot for minimally invasive surgery," in *Proceedings of the IEEE International Conference on Robotics and Automation*, Brisbane, QLD, 1147–1152.

[15] Degani, A., Choset, H., Zubiate, B., Ota, T., and Zenati, M. (2008). "Highly articulated robotic probe for minimally invasive surgery," in *Proceedings of the 30th Annual International IEEE EMBS Conference*, Vancouver, BC.

[16] OC Robotics (2008). *Snake-arm Robots Access the Inaccessible*. Bristol: OC Robotics, 92–94.

[17] Sareh, P., and Kovac, M. (2017). Mechanized creatures. *Science* 355:1379.

[18] Sklar, E., Sareh, S., Secco, E., Faragasso, A., and Althoefer, K. (2016) A non-linear model for predicting tip position of a pliable robot arm segment using bending sensor data. *Sens. Trans.* 199, 52–61.

[19] Ranzani, T., Gerboni, G., Cianchetti, M., and Menciassi, A. (2015). A bioinspired soft manipulator for minimally invasive surgery. *Bioinspir. Biomim.* 10:035008.

[20] Sareh, S., Rossiter, J. M., Conn, A. T., Drescher, K., and Goldstein, R. E. (2013). Swimming like algae: biomimetic soft artificial cilia. *J. R. Soc. Interface* 10:0666.

[21] Robertson, M. A. and Paik, J. (2017). New soft robots really suck: vacuum-powered systems empower diverse capabilities. *Sci. Robot.* 2:eaan6357.

[22] Noh, Y., Sareh, S., Wrdemann, H., Liu, H., Housden, J., Rhode, K., et al. (2015). A three-axial fiber-optic body force sensor for flexible manipulators. *IEEE Sens. J.* 99, 1641–1651.

[23] Sareh, S., Jiang, A, Faragasso, A., Noh, Y., Nanayakkara, T., Dasgupta, P., et al. (2014). "Bio-inspired tactile sensor sleeve for surgical soft manipulators," in *Proceedings of the IEEE International Conference on Robotics and Automation (ICRA 2014)*, Hong Kong.

[24] Wurdemann, H. A., Sareh, S., Shafti, A., Noh, Y., Faragasso, A., Chathuranga, D. S., et al. (2015). "Embedded electro-conductive yarn for shape sensing of soft robotic manipulators," in *Proceedings of the 37th Annual International Conference of the IEEE Engineering in Medicine and Biology Society (EMBC)*, Milano, 8026–8029.

[25] Sareh, S., Noh, Y., Ranzani, T., Würdemann, H., Liu, H., and Althoefer, K. (2015). "Modular fibre-optic shape sensor for articulated surgical instruments," in *Proceedings of the Hamlyn Symposium on Medical Robotics*, London.

[26] Croom, J. M., Rucker, D. C., Romano, J. M., and Webster, R. J. (2010). "Visual sensing of continuum robot shape using self-organizing maps," in *Proceedings of the IEEE International Conference on Robotics and Automation*, Singapore, 4591–4596.

[27] Mahvash, M., and Dupont, P. E. (2010). "Stiffness control of a continuum manipulator in contact with a soft environment," in *Proceedings of the IEEE/RSJ International Conference on Intelligent Robots and Systems,* Madrid, 863–870.

[28] Sareh, S., Noh, Y., Ranzani, T., Würdemann, H., Liu, H., and Althoefer, K. (2015). "A 7.5 mm Steiner chain fibre-optic system for multi-segment flex sensing," in *Proceedings of the IEEE/RSJ in International Conference on Intelligent Robots and Systems (IROS)*, Vancouver, BC, 2336–2341.

[29] Punning, A., Kruusmaa, M., and Abaloo, A. (2007). Surface resistance experiments with IPMC. *Sens. Act. A* 133, 200–209.

[30] Vogt, D. M., Yong-Lae, P., and Wood, R. J. (2013). Design and characterization of a soft multi-axis force sensor using embedded microfluidic channels. *Sens. J. IEEE* 13, 4056–4064.

[31] Sareh, S., Noh, Y., Li, M., Ranzani, T., Liu, H. and Althoefer, K. (2015). Macrobend optical sensing for pose measurement in soft robot arms. *Smart Mater. Struct.* 24:125024.

[32] Ryu, S., and Dupont, P. E. (2014). "FBG-based shape sensing tubes for continuum robots," in *Proceedings of the IEEE International Conference on Robotics and Automation*, Singapore, 3531–3537.

[33] Zhang, X., Max, J. J., Jiang, X., Yu, L., and Kassi, H. (2007) "Experimental investigation on optical spectral deformation of embedded FBG sensors," in *Proceedings of the SPIE 6478, Photonics Packaging, Integration, and Interconnects VII*, San Jose, CA.

[34] Polygerinos, P., Seneviratne, L. D., and Althoefer, K. (2011). Modeling of light intensity-modulated fibre-optic displacement sensors. *Act. Instr. Meas.* 60, 1408–1415.

[35] Noh, Y., Liu, H., Sareh, S., Chathuranga, D., Wurdemann, H., Rhode, K., et al. (2016). Image-based optical miniaturized three-axis force sensor for cardiac catheterization. *IEEE Sens.* 16, 7924–7932.

[36] Patrick, H., Chang, C., and Vohra, S. (1998). Long period fibre gratings for structural bend sensing. *Electron. Lett.* 34, 1773–1775.

[37] Yi, J., Zhu, X., Shen, L., Sun, B., and Jiang, L. (2010). An orthogonal curvature fibre bragg grating sensor array for shape reconstruction. *Commun. Comput. Inform. Sci.* 97, 25–31.

[38] Breedveld, P., Scheltes, J., Blom, E., and Verheij, J. (2005). A new, easily miniaturized steerable endoscope. *IEEE Eng. Med. Biol. Mag.* 26, 40–47.

8

The STIFF-FLOP Vision System

Erwin Gerz, Matthias Mende and Hubert Roth

Lehrstuhl für Regelungs- und Steuerungstechnik (RST), Department Elektrotechnik und Informatik, Fakultät IV – Naturwissenschaftlich-Technische Fakultät, Universität Siegen, Siegen, Germany

Abstract

The use of new soft robots in minimally invasive surgery offers exciting new possibilities while it generates new challenges for the technical implementation. This chapter presents methods for the detection of the STIFF-FLOP arm using visual sensing means. Based on the image information of an endoscopic camera, the visible sections are evaluated to determine the position of the manipulator. A variety of algorithms for the detection of the STIFF-FLOP arm as well as for the detection of its module connectors will be described.

A stereo camera is used to register all components in a common frame. A transformation tree is set up to refer the position of the STIFF-FLOP arm in the endoscopic camera image to the base of the STIFF-FLOP arm.

All methods have been integrated and tested in the newly developed system.

8.1 Introduction

To monitor and to control the STIFF-FLOP arm, the video data of the surgeon's endoscope is processed and evaluated. The biggest challenge lies in the reliable detection of the STIFF-FLOP arm. The nature of the object itself and the conditions in the workspace as well as the equipment available for minimally invasive surgery (MIS) leads to several restrictions in the implementation strategies.

Established methods are not applicable in the examined scenario, as these are generally either based on the detection of a known and trained outline, a clear, well-known texture or on the pronounced differences in contrast of foreground and background. None of these conditions are given here: The manipulator is flexible throughout and can change its shape and length. In addition, texture recognition is difficult to implement, since any fitted pattern is distorted significantly with increasing curvature of the arm.

The challenge here is in the detection of a flexible arm, capable of changing its shape and size during the movement, as well as the fact that the visible section of the arm can vary in the endoscope image. With the aim to develop a high-performance automated learning and recognition method, a two-step algorithm has been designed. On the one hand a texture-based pattern recognition and classification method based on Support Vector Machines (SVMs) [1–3] has been implemented [4]. The second step is the detection of optical circular markers with a modified circle detection algorithm.

8.2 Optical Tracking of the STIFF-FLOP Arm

The vision system for the tracking of the STIFF-FLOP arm consists of a 3D-tracking system (Axios, Cambar B2, Germany) and two endoscopic camera systems (Richard Wolf, Endocam 5509 and Richard Wolf, Endocam Performance HD, Germany).

In order to process the image data of the endoscopic camera, the video is streamed to ROS using a frame grabber (Intensity Pro, Blackmagic, Australia).

In the first step the endoscopic cameras will monitor and detect the STIFF-FLOP arm. In the second step the 3D-tracking system will track locators mounted on the endoscopic camera systems. Based on this data a transformation tree is set up which allows registering the position of the endoscopic camera to the base of the STIFF-FLOP arm. Figure 8.1 shows the setup for the operating room.

8.2.1 Axios Measurement System Cambar B2

The tracking system Cambar B2 is a stereo imaging system for highly accurate measurement of 3D coordinates of signalized points within a specified measurement volume. It consists of hardware components as well as software parts controlling the system.

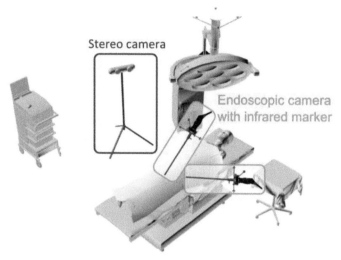

Figure 8.1 The endoscopic camera is used to detect the STIFF-FLOP arm at its destination while it is monitored by the stereo camera.

Figure 8.2 Stereo camera Axios Cambar B2, Axios 3D, Oldenburg, Germany.

The Cambar imaging system, which is shown in Figure 8.2, detects and measures passive, i.e., retroreflective points. These markers are either measured as single points or – if they fulfill a pre-defined marker geometry – as locators or rigid bodies. Measured points are classified according to their image characteristics and geometry. Afterwards, they are assigned to accuracy classes to describe and evaluate their influence on the maximum achievable measurement accuracy. The locators should meet several requirements that are summarized in Table 8.1.

Table 8.1   Specifications for the locator design

	Minimum Requirements	Recommended Specifications
Minimum number of markers	3	≥ 4
Marker shape and marker surface	Sphere, dot (flat circle) (retro-reflective)	Sphere (retro-reflective)
Marker diameter	10–12 mm (depends on distance from camera to measured object)	10 mm
Minimum distance between marker centers	Twice the minimum marker diameter	Twice the actual marker diameter
Distance of segments between markers to other segments within a locator.	> Minimum diameter	> Marker diameter
Requirements regarding rigid body geometry	Points must not be aligned in a straight line.	Points should be spread-out in space (in x-, y-, and z- direction), not be on the same plane.

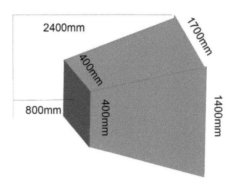

Figure 8.3   Pre-defined measurement volume of the Axios camera Cambar B2 [5].

It is advisable to use markers of the recommended specifications in order to achieve maximum accuracy. The system is capable of tracking and measuring points in a pre-defined measurement volume, which is displayed in Figure 8.3.

8.2.2 The Endoscopic Camera System

In order to find and to prove the pose and orientation of the developed STIFF-FLOP manipulator, two endoscopic cameras were ordered as shown in Figure 8.4. Each camera system consists of a light module with a xenon

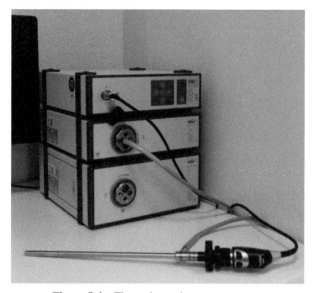

Figure 8.4   The endoscopic camera system.

light source (Light Projector 5124 and 5132, Richard Wolf, Germany) and a camera module (HD Endocam 5509 and Endocam performance HD, Richard Wolf, Germany), which is the interface to the camera on the laparoscope. The laparoscope connects the lens and the camera as well as the fiber of the light source.

The aim is to observe the STIFF-FLOP arm while it is operated by the surgeon. The laparoscopic cameras provide a video stream in full HD resolution (1920*1080, @50 Hz), in medical HD resolution (1280*1024, @50 Hz) or in HD ready resolution (1280*720, @50 Hz). To enable clinical use, all parts of the system are autoclavable (except the disposable reflectors). The video streams will be analyzed using image processing algorithms in order to detect the STIFF-FLOP arm.

8.2.3 Image Processing on Endoscopic Camera Images

The video stream of the endoscopic camera is captured with a frame grabber and streamed to ROS afterwards. Here the integrated image processing functions can be used to calibrate the camera and correct the image distortion which is displayed in Figure 8.5.

Figure 8.5   The left image shows the original image, the right shows the undistorted image.

8.2.3.1 Removal of specular reflections

Specular reflections occur if light shines on a surface. According to the laws of reflection, the reflected light beam has the same angle as the incident light beam relative to the normal of the surface. Based on the physical structure of the endoscopic cameras it is nearly impossible to get an image without specularities. These areas of overexposure lead to a corrupted filter mask, so that the algorithm could not detect the observed structure reliably. In order to remove these highlights and to reconstruct the original color, different methods were compared.

Shen and Cai [6] introduced an effective method to separate specular reflections and diffuse reflection components in multi-colored textured surfaces using a single image.

The first procedure in this method includes scanning the image to determine the minimum value of RGB components of each pixel. Each minimum value is subtracted from all three RGB components in the corresponding pixel to produce the specular-free image. After that, a threshold is added to each pixel to compensate the loss of the chromaticity that occurred because of the subtraction performed earlier to produce what Shen and Cai called the *Modified Specular Free Image*.

Another promising method was described by Miyazaki et al. [7], which is applied on single images. It does not apply any region segmentation or consider any relations between the neighbor pixels. This makes the execution time dependent only on the size of the image. The geometry of textures in the image is maintained and it does not affect the execution time either. The hue and saturation of the image do not change after the process, but the intensity does. The color changes slightly as well, but it remains similar to the color in the original image.

The method obtains the specular-free image by transforming the original image from its RGB color space into another customized color space introduced by Miayazki et al. In that color space, a filter is applied on the image data to eliminate the specular reflection components. The image is then transferred back into the RGB color space.

Both methods were implemented and tested. A comparison between the images with highlights and the result after applying both methods are shown in Figures 8.6 to 8.8.

In different tests, the method described by Miyazaki performed about 20% faster than the method described by Shen and Cai. By applying these methods on the video stream, large specular reflections were removed completely in almost all procedures. Restoring the original surface color behind large specular reflections is still problematic and as a result, those areas appear grey after the specular reflection removal was applied.

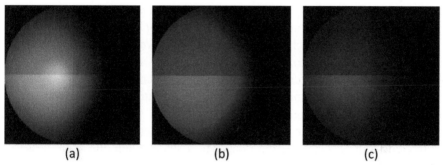

Figure 8.6 (a) Sphere with four different colors and a spot of highlights almost in the middle; (b) the result image after applying Shen and Cai's method; and (c) the result image after applying Miyazaki's method.

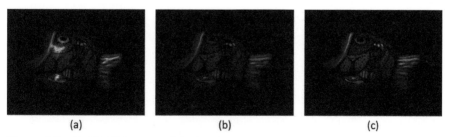

Figure 8.7 (a) Highlights on a fish; (b) the result image after applying Shen and Cai's method, and (c) the result image after applying Miyazaki's method.

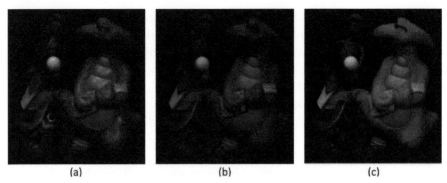

(a)   (b)   (c)

Figure 8.8   (a) Highlights on various toys; (b) the result image after applying Shen and Cai's method, and (c) the result image after applying Miyazaki's method.

8.2.3.2 Improvement of the dynamic range

Another approach that became popular in the last years is to extend the dynamic range of images. Typically this procedure requires a huge number of computations, so it is not applicable for real-time applications.

Inspired by this approach, a light-weight high dynamic range method was implemented. Hereby, one image taken with a long exposure time is combined with an image taken with a short exposure time (Figure 8.9).

After converting both images to an HSV (hue, saturation, value) color space, white areas of the lighter image are detected. These areas are also identified in the darker image, where the brightness value of the darkest pixel is subtracted from all pixels of the image. In a second step, the brightness channel of both images is added up to a new image. This newly generated image unfortunately exceeds the limits of HSV specifications, so the brightness layer has to be shrunk back into valid borders.

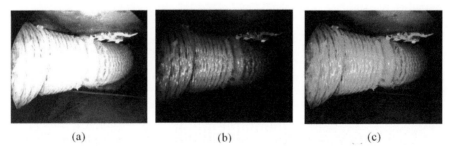

(a)   (b)   (c)

Figure 8.9   (a) Image taken with long exposure time, (b) image taken with short exposure time, and (c) the combined image with an improved dynamic range.

The resulting data works quite well as an input stream for the object-detection algorithms. Unfortunately switching from one aperture to another causes an unexpected delay, so the resulting frame rate is not satisfying.

8.2.4 Detection of the STIFF-FLOP Arm in the Camera Image using Machine Learning Algorithms

In this section, the detection of the STIFF-FLOP arm in the endoscopic camera image is described. In the first approach, it was assumed to color-code the STIFF-FLOP manipulator in an equal color and apply a color filter to detect it. Following this approach it was possible to detect the colored arm. But as soon as the color on the STIFF-FLOP arm would become contaminated the detection would fail. In the second approach, a machine learning algorithm was used to detect the STIFF-FLOP arm.

The methodology which provided the most reliable results is based on the usage of SVMs. For the use of those, a set of training samples is needed, where the searched object (i.e., the STIFF-FLOP arm), as well as the background are labeled correspondingly. The model of SVM represents the samples as points in space. They are mapped in a way that the categories are separated by a gap. For the largest gap, the recognition will provide the most stable results. This idea of data classification can be applied to images. By analyzing images, they can be classified into categories.

For this application the idea is to divide the camera image into small squares, sized about 25×25 pixels, which are analyzed and classified as background or as object (i.e., the STIFF-FLOP manipulator), labeled with the variable $y_i \in \{-1, 1\}$.

$$D = \{(x_i, y_i) \, | \, x_i \in \Re^p, \, y_i \in \{-1, 1\}\}_{i=1}^{n} \qquad (8.1)$$

where D represents the number n of quadratic parts x_i of the initial image. In Figure 8.10 a simplified visualization of the previous equation is shown, whereby the red squares represent one classification and the blue circles represent the other classification (i.e., background and object).

Assuming a dataset as observed in Figure 8.10, a hyperplane can be defined which separates both classes. The green line in Figure 8.10 represents an optimal hyperplane, which separates both classes with the maximum margin. This hyperplane will act as a classifier to recognize the object in the camera image.

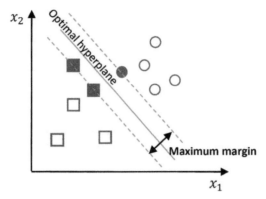

Figure 8.10 Here the different parts of the image are visualized according to their characteristics, whereby the red squares represent the first class and the blue circles the second class.

Following the procedure described by [8] and [4], a hyperplane can be constructed as the set of points $\underline{x}$ satisfying the equation:

$$\underline{w}' \cdot \underline{x} + b = 0 \tag{8.2}$$

Where $\underline{w}$ is the normal vector of the hyperplane and b is the bias. The sample points can be either found above the upper or below the lower side of the hyperplane:

$$\underline{w}' \cdot \underline{x} + b \geq 1 \quad for \quad y_i = +1$$
$$\underline{w}' \cdot \underline{x} + b \leq 1 \quad for \quad y_i = -1 \tag{8.3}$$

The samples on the upper and lower margin are the support vectors. Considering y_i they are defined as:

$$y_i(\underline{w}' \cdot \underline{x} + b) - 1 = 0 \tag{8.4}$$

In order to find the maximum margin, the distance between the support vectors on the upper and lower border has to be calculated.
    The points on the border are defined as:

$$x_+ = \frac{1-b}{\underline{w}} \quad for \quad y_i = +1$$

$$x_- = \frac{1-b}{\underline{w}} \quad for \quad y_i = -1 \tag{8.5}$$

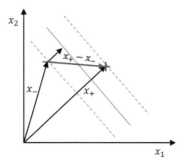

Figure 8.11 The distance between the point x_- on the lower margin and the point x_+ on the upper border is visualized.

Considering the geometry, the width is calculated by subtracting the points on the border, as displayed in Figure 8.11.

$$width = (x_+ - x_-) \cdot \frac{w}{\|w\|} = \frac{2}{\|w\|} \tag{8.6}$$

In order to maximize the margin, the minimum of w has to be determined. Therefore, the minimum of w can be substituted:

$$substitute: \quad \min \|\underline{w}\| \quad by \quad \min \tfrac{1}{2} \|\underline{w}\|^2 \tag{8.7}$$

In order to minimize $\underline{w}$ the argument is extended with the support vectors $y_i(\underline{w}' \cdot \underline{x} + b) - 1 = 0$ multiplied by the Lagrangian α:

$$L = \arg\min_{w,b} \max_{\alpha \geq 0} \left\{ \frac{1}{2} \|\underline{w}\|^2 - \sum_{i=1}^{n} \alpha_i \left[y_i(\underline{w}' \cdot \underline{x} + b) - 1 \right] \right\} \tag{8.8}$$

The minimum is found by setting the first derivative to zero. The resulting $\underline{w}$ is found as a linear combination of the samples with $\underline{x}_i$ as support vector.

$$\underline{w} = \sum_{i=1}^{n} \alpha_i y_i \underline{x}_i$$
$$\sum_{i=1}^{n} \alpha_i y_i = 0 \tag{8.9}$$

If the classes cannot be separated linearly, the argument can be extended by a polynomial or a radial basis function, allowing a more complex separation [2]. For this application, the best results were obtained with a polynomial approach of third order.

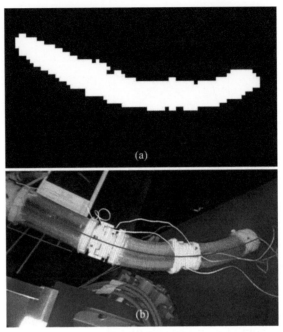

Figure 8.12   (a) Detected parts of the STIFF-FLOP arm using the prior calculated hyperplane (b) corresponding camera image with the highlighted centerline (red) and perpendicular to it the diameter is calculated (violet).

After the hyperplane is identified, it can be used for a real-time application providing the possibility to process the acquired camera images directly. During the real-time application, the same idea is pursued. The camera image is grained into small parts with the same size that was used during the learning process. Afterwards these small parts are classified using the calculated hyperplane. As result, the detected contour is visualized in Figure 8.12.

The next step is to calculate the spatial position of the manipulator in the camera coordinates. Therefore, the center line of the detected STIFF-FLOP arm is determined and the diameter is calculated accordingly. Based on the information, the planar coordinates (i.e., x and y) can be extracted from the pixel coordinates and the distance of the STIFF-FLOP arm to the camera's optical center is calculated based on the determined diameter.

In Figure 8.12, the usage of the implemented machine learning STIFF-FLOP arm detection algorithm is demonstrated, allowing an estimation of the position of the STIFF-FLOP arm in camera coordinates.

8.2.5 Detection of the Module Connection Points of the STIFF-FLOP Arm

To achieve a reliable and robust tracking result, in particular when the STIFF-FLOP arm is being maneuvered in front of unknown and challenging backgrounds inside the abdominal area, unambiguous green ring markers are placed on the STIFF-FLOP manipulator at distinct locations as shown in Figure 8.13.

The color is chosen due to the high contrast with respect to the expected color scheme (mainly in the red spectrum) during surgical interventions inside the abdomen. The ring markers are fixed on the force/torque sensor structures that are placed between segments and at the end of the STIFF-FLOP manipulator.

The detection of these circular markers has several advantages. Firstly, the figure remains almost distortion-free with sufficiently narrow markings. On the other hand, a possibly occurring system-related radial extension of the arm solely affects the accuracy of the distance measurement between the camera plane and the central axis of the manipulator. The presented method is also robust against noise or smaller highlights, as long as the contour of the ring is not interrupted over larger sections.

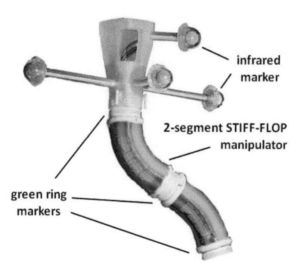

infrared marker

2-segment STIFF-FLOP manipulator

green ring markers

Figure 8.13 Two-segment STIFF-FLOP manipulator with green marker rings placed at the base of the STIFF-FLOP arm and at the module connectors.

Reproducible results were achieved by the application of green ring markers as displayed in Figure 8.13. The idea is based on an algorithm for camera calibration with two arbitrary coplanar circles [9].

The detection process is visualized in Figure 8.14. As a first step, a mask containing all the green colored areas of the video image is generated (2). This mask is the source for a standard contour detection algorithm. The detected contours are filtered based on a set of rules eliminating noise and other unsuitable objects so that only sections of ring-like objects of a suitable size should reach the next step (3).

The remaining contours are split into a concave (4, blue curve) and a convex (4, red curve) part. Each of them is now treated like the visible part

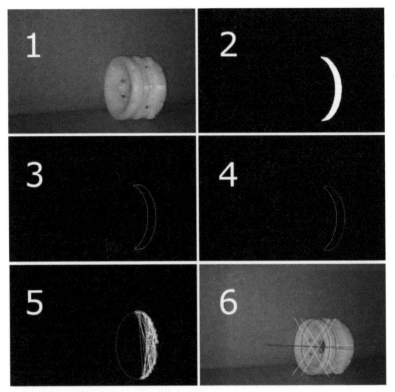

Figure 8.14   Ring marker detection sequence, (1) shows the camera image of the ring marker on the module connector, (2) visualizes the detected mask of the ring marker, (3) highlights the contour of the mask, in (4) this contour is divided into a concave (blue) and in a convex (red) part of an ellipse. In (5) the RANSAC approach to fit an ellipse into the detected shape is demonstrated, while (6) shows the finally determined ellipses with their center positions.

of a closed elliptical shape. Unfortunately, these contours cannot be used directly to determine the parameters of an ellipse, because they contain too much noise and therefore the ellipse parameters are oscillating in a wide range.

A RANSAC [10] approach seems suitable to improve the results, where 10% of each contour's points are selected and fed into a least-square-error algorithm to define the ellipse's parameters. The resulting ellipse is compared with the contour as a whole and the average error is evaluated. If the ellipse fits well, the parameters are stored. If not, another set of points is generated as a basis for the parameter calculation. The results can be seen in (5).

The determined ellipses can be seen as projections of circles. Knowing the circles' diameters, the calculation of those circles' poses in space is possible (6). As an ellipse is effectively a projection of a circle, the position of the midpoint can be determined which is equivalent with the center position of the force/torque sensor. The green ring markers are detected in the camera's coordinate system and are transformed in the manipulator's base system.

8.2.6 Registration of the Endoscopic Camera Image to the STIFF-FLOP Arm

In order to connect the image plane of the endoscopic camera with the world coordinates of the STIFF-FLOP arm, both the endoscopic camera and the base of the STIFF-FLOP arm were equipped with infrared markers, which are tracked by an optical 3D localization system (Axios 3D, Cambar B2, Germany). Using this, a transformation tree was determined to register the image plane of the endoscopic camera in the world coordinates of the STIFF-FLOP arm.

The ring markers are detected in the camera's coordinate system and have to be transformed in the manipulator's base system. Therefore the camera coordinate system has to be integrated into the transformation tree of the whole system by using the output of the Cambar B2 tracking system. An overview of the complete transformation tree can be seen in Figure 8.15.

The transformation tree shows that the world coordinates of the STIFF-FLOP arm are connected to the STIFF-FLOP base (SF_BASE). Through the stereo camera, the transformation of the endoscopic camera (ENDO) is known and connected to the STIFF-FLOP base.

The position of the detected ring markers are known in the frame of the endoscopic camera and are connected below in the transformation

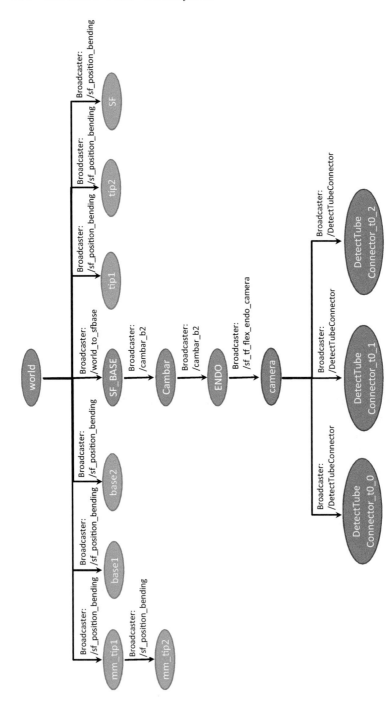

Figure 8.15 Illustration of the transformation tree of the STIFF-FLOP setup connected through ROS. The transformations generated by the optical 3D Tracking system are marked blue, and the transformations resulting from the vision system's output are marked red. The ovals describe the different Frames. The black arrows show through which Broadcasters the Frames are connected.

tree. Thereby the transformation of each detected position relative to the STIFF-FLOP base is known.

8.3 Integration and Validation of the Implemented Methods

Both the calculated centerline and the identified module connection points were determined in the image plane of the endoscopic camera. Using the transformation tree, those positions were transferred to the world coordinate frame of the STIFF-FLOP arm.

The vision system takes over the function of a second-level observation system – therefore an independent ROS node with an alert function was implemented. This node compares the connector's frames sent by the control algorithm with the detected frames published by the vision system and calculates a normalized warning level (0.0–1.0, the absolute value for inacceptable displacements of the manipulator can be reconfigured dynamically). The integration of the vision system into the open loop control is visualized in Figure 8.16.

The position of the used coordinate systems is visualized in the 3D visualization tool for ROS (rviz) (Figure 8.16). The vision system supplies a second-level observation system and optimizes the position data obtained

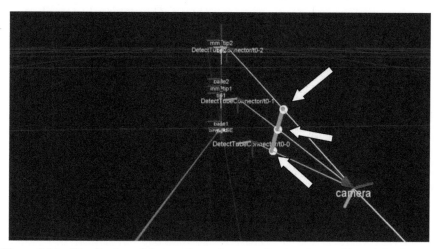

Figure 8.16 Visualization of the transformation tree, showing the matching of the detected positions of the ring markers (white bulbs) with the position provided by the control system (green line connections between ring connections).

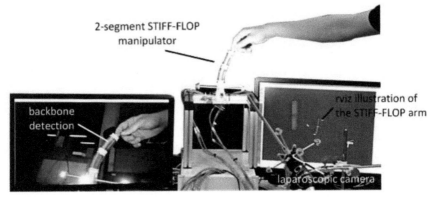

Figure 8.17   Experimental setup, demonstrating the integration of the implemented method, by detecting a deviation between the position detected by the vision system (real arm in the center and the corresponding image on the left screen) and the position of the model of the STIFF-FLOP arm (right screen) which is visualized using the ROS visualization tool (rviz). The deviation is highlighted by coloring the arm in red (right screen).

from calculations using force/torque sensor data and bending sensor information. Figure 8.17 shows the experimental setup: A 2-segment STIFF-FLOP manipulator is mounted on a rack with three green ring markers attached to the base of the robot arm and at the tip of each segment. On the left screen, the video screen of the endoscopic camera is shown. The previously described image processing algorithm detects the markers. A spline interpolation (red) exposes the backbone of the manipulator. The right screen shows this data within the 3D visualization tool.

Both algorithms have been verified in an experimental setup, which is shown in Figure 8.17. The manipulator is located in the middle; the left monitor shows the recorded video stream. The right monitor shows a simulation of the manipulator based upon the model that is used for controlling the manipulator.

The detected position of the connectors is compared to the positions of the control model. If a deviation – which might be caused by external forces – is detected, the deformation is recognized by the optical tracking system and an alarm signal is displayed by coloring the simulated arm in red.

A comparison of the detected positions of the STIFF-FLOP arm and the location given by the control model is possible after transforming all data into a common frame. Afterwards an evaluation of the detection error is possible and the control parameters can be adjusted accordingly.

8.4 Conclusion

The detection of these circular markers has several advantages. Firstly, the figure remains almost distortion-free with sufficiently narrow markings; on the other hand, a possibly occurring system-related radial extension of the arm solely affects the accuracy of the distance measurement between the camera plane and the central axis of the manipulator. The presented method is also robust against noise or smaller highlights, as long as the contour of the ring is not interrupted over larger sections.

The detection of the STIFF-FLOP arm with the implemented SVM method allows a trainable approach to adapt the detection of the STIFF-FLOP arm to the given scenario.

The redundancy of the two methods used (SVM and Ring Marker Detection) essentially offers two advantages: First, it allows a plausibility check of the detected marker positions; on the other hand the calculation of a continuous center line is possible. In addition, this center line can be used for a simple collision detection, if 3D models of the working space are provided.

Acknowledgements

The work described in this paper is funded by the Seventh Framework Programme of the European Commission under grant agreement 287728 in the framework of EU project STIFF-FLOP.

References

[1] Vapnik, W., and Chervonenkis, A. (1974). *Theory of Pattern Recognition* [in Russian] (USSR: Nauka).

[2] Boser, B. E., Guyon, I. M., and Vapnik, V. N. (1992). "A training algorithm for optimal margin classifiers," in *Proceedings of the Fifth Annual Workshop on Computational Learning Theory*, Pittsburgh, PA, 144–152.

[3] Winston, P. H. *6.034 Artificial Intelligence: Support Vector Machines, Massachusetts Institute of Technology: MIT OpenCourseWare*. Available at: http://ocw.mit.edu/courses/electrical-engineering-and-computer-science/6-034-artificial-intelligence-fall-2010

[4] Chang, C.-C., and Lin, C.-J. (2011). LIBSVM: a library for support vector machines. *ACM Trans. Intell. Syst. Technol.* 2:27.

[5] Axios 3D services (2014). *Optisches Messsystem CamBar B2 C8, Datasheet, Mai 2014.*

[6] Shen, H. L., and Cai, Q. Y. (2009). Simple and efficient method for specularity removal in an image. *Appl. Opt.* 48, 2711–2719.

[7] Miyazaki, D. (2013). Specular-Free Image. Hiroshima: Hiroshima City University. Available at: http://www.cg.info.hiroshima-cu.ac.jp/~miyazaki/knowledge/teche40.html

[8] Ben-Hur, A., and Weston, J. (2010). *A User's Guide to Support Vector Machines, Methods in Molecular Biology.* Clifton, NJ: Humana Press. Available at: https://www.researchgate.net/publication/41896604

[9] Chen, Q., Wu, H., and Wada, T. (2004). *Camera calibration with two arbitrary coplanar circles. Computer Vision—ECCV 2004. Lecture Notes in Computer Science*, Vol. 3023 (Berlin, Heidelberg: Springer), 521–532.

[10] Fischler M. A., and Bolles, R. C. (1981). Random sample consensus: a paradigm for model fitting with applications to image analysis and automated cartography. *Comm. ACM.* 24, 381–395.

PART III

Control, Kinematics and Navigation

9

Inverse Kinematics Methods for Flexible Arm Control

**Anthony Remazeilles, Asier Fernandez Iribar
and Alfonso Dominguez Garcia**

Tecnalia Research & Innovation, Donostia – San Sebastián, Spain

Abstract

Traditional minimally invasive robots provide to the surgeon an interface for controlling the tip of the endoscopic arm in Cartesian space. We proposed therefore a similar interface for the STIFF-FLOP robot. The direct control of the tip pose was provided by an inverse kinematics component, computing the appropriate STIFF-FLOP robot configuration. Due to the flexibility of the arm modules, we have organized the inverse kinematics into two layers. The first one handles the inverse kinematics in a generic way. It is based on a numerical estimation of the robot. This layer is generic in the sense that it can incorporate any module representation, as long as the module representation provides a forward kinematics mechanism. The second layer concerns the kinematic modeling of the flexible modules, and has to provide forward kinematics functionalities for the upper model. Instead of the standard constant curvature parameters, we are proposing two other representations, one using each module tip position, and the other one directly using the chamber lengths. The flexible modules are connected to a robotic arm through a rigid rod, to extend the operational space of the system. The robotic arm pose is encoded with an adaptation of the spherical coordinate system to ensure that the rod entering the human body respects the single insertion point constraint. By defining a forward kinematics for the rod pose, the external robot end effector is implicitly embedded into the general inverse kinematics scheme, so that the estimation of the flexible modules' configurations and the pose of the robot end-effector are all computed together to follow the motion requests provided by the surgeon.

9.1 Introduction

9.1.1 On the Inverse Kinematics Problem for Continuum Robots

The design of a user interface for controlling a (piecewise) continuum robot like the STIFF-FLOP arm is challenging, since such robot presents many more control parameters than can be provided through a traditional haptic device such as the one presented in Figure 9.1 (for description of haptic device refer to Chapter 16). Such conventional surgeon interfaces are used for providing the desired six DOF pose of the surgical tool tip, while the actuation space of the robot is much higher. In the case of the STIFF-FLOP robot, each flexible module is controlled with three parameters (pressure in each chamber), so that the flexible arm requires defining $3n$ parameters, n being the number of modules used within the arm. If the flexible arm is mounted onto a rigid robot to extend the actuation envelope (Figure 9.2 and section 9.2.4), the positioning of this additional robot needs also to be controlled.

This problem is classical in robotics theory, and is related to *Inverse Kinematics*. As stated in [1] it consists of finding all the geometric parameters of the manipulator given the desired position and orientation of the end-effector. Considering our robot's specificities, the work of Webster et al. is particularly relevant since it provides key kinematics models for piece-wise constant curvature continuum robots [2]. Under the piecewise constant curvature assumption, the modeling of such robotic structures can be seen

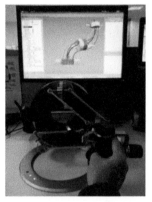

Figure 9.1   Example of a haptic device used to receive the motion request for the robotic tool (Omega 7 from Force Dimension[1]). For description refer to Chapter 16.

---

[1]http://www.forcedimension.com

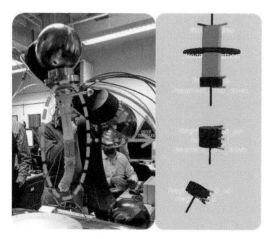

Figure 9.2 Real STIFF-FLOP system (left) vs. the simulated version used in this chapter (right). A rigid section (green on the right) is connecting the set of flexible modules to the SCHUNK arm. The connection with the first module is what we call the base in this chapter. On the right picture, only the rigid components are displayed. All modules are started and finished by a rigid section (respectively gray and purple). Once the flexible modules are inserted into the body, the motion of the SCHUNK arm must maintain a unique insertion location. In simulations involving the motion of the base, the red disc (top right) represents that constraint.

as a composition of robot-specific mapping and robot-independent mapping (Figure 9.3). The forward kinematics consists of the operations described on the upper side of the figure (going from the actuator space to the task space), while the inverse kinematics focuses on the lower operations (going from the task space to the actuator space).

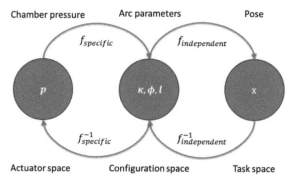

Figure 9.3   Spaces and mappings of constant curvature robots, as described in [2].

The use of arc parameters in the configuration space (κ, ϕ, l, respectively for the curvature, the angle of the plane containing the arc, and the arc length) to represent the state of a continuum flexible module respecting the constant curvature model makes sense. Nevertheless, the constant curvature model may lead to some kinematic singularities, e.g., when a flexible module is purely extended so that its curvature κ is equal to 0. Furthermore, it does not allow taking into account the potential deviations from the theoretical model, due, for example, to the external forces acting on each module. In the latter case, a Jacobian formulation in closed form of the independent mapping, as proposed in [2], may not be easily obtained.

Following the suggestions of our colleagues in [3], we propose to relax the central role of the constant curvature parameter in modeling, and propose a generic model of the configuration space. We combine this with a numerical computation of the Jacobian. The advantage of using a Jacobian estimated numerically is that it allows considering any configuration space that provides a forward kinematics mapping. This approach is also used to deduce, simultaneous to the estimation of the appropriate configuration of the flexible modules, the motion of the robotic arm holding the STIFF-FLOP flexible modules, while respecting the single insertion point constraint that is described in the following section.

9.1.2 Single Insertion Point Constraint in Minimally Invasive Surgery

In the context of robotic surgery and minimally invasive surgery (MIS), the main purpose of using a tool composed of flexible modules is to permit accessing spaces that cannot be directly reached by a rigid structure, limiting thus the multiplication of insertion points and the number of needed incisions into the human body. Nevertheless, it is still preferable to mount the flexible modules onto a standard robotic arm to augment the reachable space, and only employ the bending properties of the modules when complex displacements are required. Naturally, the combination of the robotic arm with the flexible modules increases the number of actuation parameters, and an automatic control of that robotic arm is required to maintain the classical interface used by surgeons.

Any robotic system involved in MIS must respect the single insertion constraint. The instrument held by the robot is inserted into the human body through a trocar at the incision point, and the trocar position needs to be maintained throughout the surgery. This constraint can be solved by the design of dedicated mechanical structures that inherently respect the remote center of motion, such as the well-known Da Vinci [4], or endoscope maintainers directly placed onto the body, such as [5]. Another approach consists in equipping the robot with passive joints at the wrist so that the instrument naturally rotates around the fulcrum point [6, 7], but backlash may appear in some configurations leading to a lack of control of the instrument motion.

The use of more regular six-DOF robots can be seen as a more versatile solution involving a light robotic system that facilitates its displacement during surgery (in particular when being moved from a trocar to another [8]). The insertion point constraint is ensured by dedicated controllers, leading to a programmable remote center of motion. In [8], the trocar constraint is modeled as a variable point along a given robot link, and is considered as an additional joint added to the arm configuration space ones during the inverse kinematics process. In [9], the task space is extended with the trocar position to produce movements that restrict its motion. In [10], a force sensor is placed at the end effector of the arm to adjust the lateral motion of the arm for limiting the forces applied at the trocar site. Other works based on visual servoing directly adjust the interaction matrix linking the motions of a camera to the image point motion for taking into account the trocar constraint that reduce the displacement of the endoscope [11, 12].

In the context of programmable remote center of motion, the modeling of the trocar constraint with spherical coordinates seems to be particularly appropriate since, *per se*, spherical coordinates can only describe directions going through the origin (which is placed at the trocar frame, as illustrated in Figure 9.4) [10, 13, 14].

We propose in this chapter to embed the robotic arm in the inverse kinematics framework by inserting a component defining the location of the rigid connector between the end effector of the arm and the first flexible module. The pose of the rigid connector is defined with spherical coordinates in which the origin frame is placed at the insertion point of the trocar.

9.1.3 Contributions Presented

The present chapter presents how we propose to simultaneously estimate the appropriate configuration of the flexible modules together with the needed robotic arm end-effector pose for reaching a desired tip location provided by the surgeon. The next section describes the generic inverse kinematics framework that is used. It is generic in the sense that it is independent of the

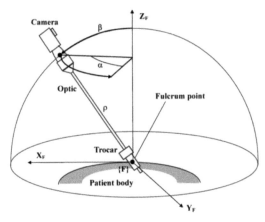

Figure 9.4   Spherical coordinates used for expressing the pose of an endoscope with respect to the insertion point, as proposed in [13].

configuration space selected for representing the state of the flexible modules (several configuration spaces are considered in section 9.2.3), and it can be applied for any number of flexible modules. Furthermore, it also enables to consider the external robotic arm onto which the flexible modules are mounted, and to deduce the appropriate pose of this arm end-effector while ensuring the trocar constraint. Section 9.2.4 will show how the traditional spherical coordinates can be adjusted to better fit with the surgical context. We will then show in section 9.2.5 that the redundancy of our global system can be used to consider secondary tasks improving the behavior of the flexible modules in the patient body. Finally several simulations will be detailed in section 9.3 to demonstrate the validity of the contributions proposed.

9.2 Inverse Kinematics Framework

9.2.1 General Framework

We consider that we know the current pose of the STIFF-FLOP tip with respect to a world frame W. We note it:

$$^W M_t = \begin{bmatrix} ^W R_t & ^W t_t \\ 0 & 1 \end{bmatrix},\tag{9.1}$$

and we consider that a desired tip pose is defined, as $^W M_t^*$. In the spirit of the formalization proposed in [3], the control of the complete STIFF-FLOP system relies on the definition of a task function e representing the error between the current tip pose and the desired one. The task error is defined as:

$$e = [\mathbf{\Delta}_t, \theta u]^\mathrm{T},$$ (9.2)

where the first three entries $\mathbf{\Delta}_t$ are related to the difference between the current tool tip position and the desired one:

$$\mathbf{\Delta}_t = {}^W t_t - {}^W t_t^*,$$ (9.3)

and θu is the axis-angle representation of the orientation difference between the current tip frame and the desired one. In [3], the orientation error was only considering the x and y components of the rotation vector. To be more general, we extended the error model to contain as well the rotation around the z-axis, so that the error model represents the complete pose of the tip.

The task function variation can be related to the system parameterization q:

$$\frac{de}{dt} = \frac{\partial e}{\partial q}\frac{dq}{dt} = \mathbf{J}\frac{dq}{dt},$$ (9.4)

where $\mathbf{J}$ is the Jacobian that links the evolution of the task function to the variation of the variables contained in q, so that:

$$\dot{q} = \mathbf{J}^+\dot{e}$$ (9.5)

The usual models used for the task error evolution $\dot{e}$ along time are either affine or exponential. In an affine model $\dot{e} = -\lambda$, while in a model with an exponential decrease $\dot{e} = -\lambda e$.

In the context of the STIFF-FLOP project, we proposed to compute the Jacobian numerically. The advantage of such an approach is that we can seamlessly investigate new module models, by adjusting accordingly the related parameters in q, and by providing the related forward kinematics for each piecewise component. The numerical estimation of the Jacobian is based on its structure:

$$\mathbf{J} = \begin{bmatrix} \frac{\partial e}{\partial q_0} & \cdots & \frac{\partial e}{\partial q_n} \end{bmatrix},$$ (9.6)

So that each column of the Jacobian can be estimated in the following way:

$$\frac{\partial e}{\partial q_i} = \frac{e\left(q+\delta q_i\right) - e\left(q\right)}{\delta}$$ (9.7)

9.2.2 Application to the STIFF-FLOP Structure

The generic formulation of the structure configuration q is defined as a stacking of the configuration of each sub-component constituting the STIFF-FLOP robot: a configuration q_b of the base of the flexible modules (the position of which is adjusted using the standard robotic arm holders), and n sub-configurations specification for each of the n flexible modules considered (three are presented in Figure 9.5, but the principle is generic).

In the above example, the pose of the tip of the STIFF-FLOP structure can be obtained by the composition of the pose of each successive component. Following the notation used in the previous section, we can write:

$$^wM_t = {}^wM_b.{}^bM_1.{}^1M_2.{}^2M_3 \tag{9.8}$$

The first transform is the definition of the pose of the base of the flexible modules, handled with the standard robotic arm. The following transforms are obtained by applying the forward kinematics mapping from the related module configuration q_i (and including the connecting rigid sections that are skipped for notation simplicities).

As previously stated, the choice of the parameterization format of each module and of the base is transparent for the inverse kinematics model based on a numerical estimation of the Jacobian. It is only necessary to have a

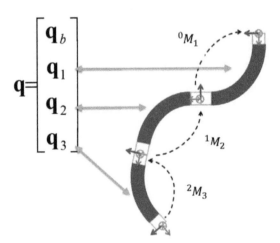

Figure 9.5   Relation between the parameterization of the STIFF-FLOP structure and the tip pose. q_b refers to the configuration of the base of the flexible structure, which is controlled through the robotic holder not depicted here. Each module is equipped with two rigid connections to allow inter-module attachments.

forward kinematic mapping from this configuration space, to provide by composition the expression of the tip location with respect to a reference world frame. In particular, the expression of the base location q_b is different from the one used for the flexible modules q_i, since they correspond to components which are totally different. But they can be stacked in a common vector to produce the global configuration space of the STIFF-FLOP robot.

9.2.3 Configuration Space of the Flexible Modules

In [2], and as illustrated in Figure 9.3, the configuration space chosen for the flexible modules is directly the constant curvature parameterization, so that each module will be represented by the feature $q_i = [\kappa_i, \phi_i, l_i]^{\top}$. Such parameterization suffers from a representation singularity when a module is purely extended: the curvature is null, and the plane angle ϕ can have any value.

In [3], it is proposed to use the position of the tip of each flexible module with respect to its base (noted Q_i). This means that the parameterization q is defined as (omitting the base pose):

$$q = [Q_0, \ldots, Q_{k-1}]^{\top}. \tag{9.9}$$

The forward kinematics of a module from such configuration requires the computation of the module orientation R_i, which is under the constant curvature assumption [3], assuming $Q_i = [x_i, y_i, z_i]^{\top}$:

$$R_i = \frac{1}{Q_i^{\top} Q_i} \begin{bmatrix} -x^2 + y^2 + z^2 & -2xy & 2xz \\ -2xy & x^2 + z^2 - y^2 & 2yz \\ -2xz & -2yz & z^2 - x^2 - y^2 \end{bmatrix} \tag{9.10}$$

As stated, other models can be considered in this generic inverse kinematics framework. For instance, instead of the tip position, one can use the length of the pressured chambers, *i.e.*, $q_i = [l_1, l_2, l_3]^{\top}$. In that case, the forward kinematics requires computing the constant curvature parameters of each module (noting here the constant curvature with the bending angle α, the orientation angle β and the chamber length L as in [15]):

$$L = \frac{1}{3}(l_1 + l_2 + l_3)$$

$$\alpha = a \tan 2 \left(\sqrt{3}(l_3 - l_2), l_2 + l_3 - 2l_1 \right)$$

$$\beta = \frac{2\sqrt{l_1^2 + l_2^2 + l_3^2 - l_1 l_2 - l_1 l_3 - l_2 l_3}}{3r}$$

with r being the radial distance of the chamber from the module center. From the constant curvature model, the pose of the tip of a module with respect to its base is:

$$^b t_t = \frac{L}{\beta} \begin{bmatrix} \cos \alpha \ (1- \cos \beta \ ) \\ \sin \alpha \ (1- \cos \beta \ ) \\ \cos \beta \end{bmatrix}. \tag{9.11}$$

And the orientation is obtained through:

$$^b R_t = \begin{bmatrix} \cos^2\alpha \ (\cos \beta - 1) + 1 & \sin \ \alpha \cos \ \alpha \ (\cos \beta - 1) & \cos \ \alpha \sin \beta \\ \sin \ \alpha \cos \ \alpha \ (\cos \beta - 1) & \cos^2\alpha \ (1 - \cos \beta \ ) + \cos \beta & \sin \ \alpha \sin \beta \\ -\cos \ \alpha \sin \beta & -\sin \ \alpha \sin \beta & \cos \beta \end{bmatrix} \tag{9.12}$$

The two previous module models rely on the constant curvature assumption. Even though it is not demonstrated here, it is possible to use models relaxing that hypothesis, as it is done in the work of [16]. In this beam theory-based model, the forces measured at each module junction are taken into account to better estimate the deformation of each flexible module (which in this case does not follow the constant curvature hypothesis). With this model, the configuration space of q would be equivalent to the actuator space, i.e., the pressures being applied in the chamber. The forward kinematic model proposed in [16] is compatible with our numerical estimation of the Jacobian. The only difference is that forward kinematics is computed for all flexible modules simultaneously, while in the two previous models introduced here (based on module tip poses and based on chamber lengths), forward kinematics is computed per module independently and then composed.

9.2.4 STIFF-FLOP Base Motion with Single Insertion Point Constraint

As already stated in section 9.1.2, mounting the flexible manipulator onto a standard robotic arm enables extending the reachable workspace, while focusing the use of the bending capabilities of the flexible manipulator to areas not reachable to rigid structures with a linear motion. In the context of the STIFF-FLOP project, the flexible manipulator was fixed with a rigid rod at the end-effector of a SCHUNK LWA (as illustrated in Figure 9.2). In the rest of the section, we define the position of the SCHUNK arm by the location of the base of the first flexible module (i.e., just after the green rigid section presented in Figures 9.2 and 9.6). We assume that the standard robotic arm

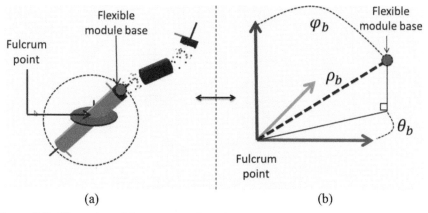

Figure 9.6 Illustration of the representation of the flexible STIFF-FLOP arm fixed onto a movable structure. (a) The flexible modules are mounted onto a rigid rod (green cylinder), which must respect a fulcrum point (at the center of the red disc). The red dot indicates the tip of the rigid rod that is expressed using spherical coordinates. (b) Spherical coordinate model used to express the pose of the rigid rod tip with respect to an origin placed at the fulcrum point. The red-green-blue vectors represent the reference frame.

is equipped with inverse kinematics means to directly control the pose of the end-effector.

In Figure 9.6, the green cylinder emulates the rigid component onto which the base of the first flexible module is mounted. Again, this rigid component is mounted itself at the end effector of a robotic arm (which would be placed at the bottom left end of the green cylinder on Figure 9.6) providing motion capabilities to the base of the STIFF-FLOP arm. The red disc emulates the insertion point (i.e., the trocar port). Any generated motion of the global system should go through that point. Using spherical coordinates for the robot base, the fulcrum constraint can easily be computed and ensured.

It is straightforward to take the base motion into account within the inverse kinematic model previously defined. Indeed, one can define a specific feature, **B**, gathering the needed parameters to define the base pose with respect to a world frame. If the base would be totally free of motion (the so-called *free-flying base*), then the base model could be chosen to be a vector of six components:

$$\mathbf{q_b} = \begin{bmatrix} {}^W t_b \\ \text{euler} \left({}^W \mathbf{R}_b \right) \end{bmatrix}, \tag{9.13}$$

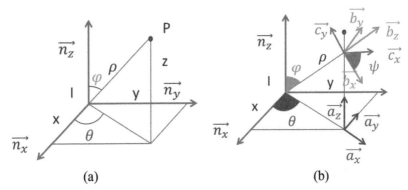

(a)   (b)

Figure 9.7   Spherical coordinates (a) parameters ρ, θ, φ stand respectively for the radial distance, the azimuthal angle and the polar angle. (b) Frames generated by a spherical parameterization. Green: standard frame derived from the coordinates. Gray: adding a fourth dimension to model the orientation along the z-axis.

where the Euler function provides the three Euler angles related to a rotation matrix. In that case, the whole joint vector is defined as:

$$q = \left[q_b^{\top}, Q_0, \ \dots \ Q_{k-1} \right]^{\top},\tag{9.14}$$

and the inverse kinematics is handled following the same methodology as previously defined.

Nevertheless, the robotic arm motion must be constrained to make sure that the single point insertion constraint is respected. As illustrated in Figures 9.6 and 9.7, this insertion point denoted by I (located on the abdominal wall) acts as a pivot point or fulcrum, and spherical coordinates can be used to specify the position of the rigid rod P with respect to the reference frame located at I.

The use of spherical coordinates $[\rho, \ \theta, \ \varphi]^{\top}$ is a convenient way to restrict the possible rotations of a frame attached to P to the ones that can be generated by the pivot point I. As can be seen in Figure 9.7(b), the two angles θ and φ define the orientation of the green frame with respect to the blue reference one. Note that the green frame has its z-axis aligned with the vector $\overrightarrow{IP}$.

The related frame orientation can be deduced from the spherical coordinates:

$$^{I}\mathbf{R}_{B} = \begin{bmatrix} \cos\theta\cos\varphi & -\sin\theta & \cos\theta\sin\varphi \\ \sin\theta\cos\varphi & \cos\theta & \sin\theta\sin\varphi \\ -\sin\varphi & 0 & \cos\varphi \end{bmatrix}\tag{9.15}$$

Considering that the rotation matrix column refers to the expression of $\vec{b_z}$ in the reference frame **I**, the position of point **P** in frame **I** is directly deduced as:

$$^I t_B = \rho \vec{b_z} = \rho \begin{bmatrix} \cos\theta \ \sin\varphi \\ \cos\theta \ \sin\varphi \\ \cos\varphi \end{bmatrix}. \tag{9.16}$$

Thus, the configuration of the base can be defined by the spherical coordinates (i.e., $q_b = [\rho, \theta, \varphi]^T$), and the two previous relations correspond to the forward kinematics relations needed for the Jacobian estimation.

In Figure 9.7(b), another dimension is added to the spherical coordinates to model the rotation along the z-axis (gray frame). In that case, the orientation of the obtained frame can be defined as:

$$^I R_C = {}^I R_B R_\psi$$

$$= \begin{bmatrix} \cos\theta\cos\varphi & -\sin\theta & \cos\theta\sin\varphi \\ \sin\theta\cos\varphi & \cos\theta & \sin\theta\sin\varphi \\ -\sin\varphi & 0 & \cos\varphi \end{bmatrix} \begin{bmatrix} \cos\psi & -\sin\psi & 0 \\ \sin\psi & \cos\psi & 0 \\ 0 & 0 & 1 \end{bmatrix}$$

$$\tag{9.17}$$

$$^I R_C = \begin{bmatrix} c\theta\, c\varphi\, c\psi - s\theta s\psi & -c\theta\, c\varphi\, s\psi - s\theta\, c\psi & c\theta\, s\varphi \\ s\theta\, c\varphi\, c\psi + c\theta s\psi & -s\theta\, c\varphi\, s\psi + c\theta\, c\psi & s\theta\, s\varphi \\ -s\varphi\, c\psi & -s\varphi\, s\psi & c\varphi \end{bmatrix}$$

with c, s standing respectively for cos and sin operators.

As expected, the position of the point **P** (third column) in the reference frame remains unchanged.

Once more, this model, $q_b = [\rho, \theta, \varphi, \psi]^T$, can be included within our inverse kinematics, considering the forward kinematics provided by the relations previously introduced.

In [17], we have proposed an adjustment of the traditional spherical coordinates to get a better behavior in the context of the trocar constraint within MIS. Looking at the rotation described in Equation (9.17), we can note that the orientation induced by the extended spherical coordinates is similar to Euler angles in the ZYZ configuration of the rotations. It turns out that the system cannot thus directly handle pure rotations around the x-axis, but has to combine rotations around the z- and y-axes to produce them. The proposed adjustment consists of changing the reference angle to switch to a XYZ model instead. As stated in [17], it also enables the elimination of a representation singularity occurring with the traditional model when the green rigid component is aligned with the reference z-axis of the insertion frame. Readers may refer to [17] for further details on this model adjustment.

9.2.5 Secondary Tasks through Redundancy

Redundant robotic systems permit considering additional tasks while satisfying a main task. Our robotic system, controlled by $4+3n$ parameters (considering the mobile base, and n flexible modules), operates in a 6-dimensional task space and is, thus, highly redundant. A secondary task can thus be applied in the null space of the first task. This null space can be derived from the Jacobian matrix:

$$\mathbf{P_e} = \mathbf{I_6} - \mathbf{J}^+\mathbf{J}. \qquad (9.18)$$

Any secondary task projected onto this null space can thus be taken into account without affecting the main task completion. The term "secondary" is used considering that the main task remains the tip positioning, and the second one is only applied onto the null-space left by the primary task, i.e., the secondary task is applied only as long as it does not affect the main one.

The null space can be used to try maintaining the modules around their mean length (to avoid too large an extension in a single module) for instance, or to limit the contact with the environment which could be sensed, for example, by tactile sensors.

9.2.5.1 Control of the chamber lengths

The redundancy is frequently used for joint limit avoidance. In our case, the joint limits are related to the rest of the length of the module (since we cannot reduce the chamber length any further) and the maximum chamber extension. We can then define a secondary task to maintain each module, as much as possible, at its mean size, $\overline{L}$. $\overline{L}$ can be defined for example as $\overline{L} = L_0 \left(1 + \frac{\alpha}{2}\right)$ where L_0 is the length when the module is at rest and α is an elongation factor. We can thus define another task function W_2 relating the distance of the modules to their mean size:

$$W_2 = \frac{\sum |L_i - \overline{L}|}{n\overline{L}}. \qquad (9.19)$$

Employing a gradient projection approach, we compute the gradient of the original task function W_2,

$$\nabla_{\mathbf{q}} W_2 = \frac{\partial W_2}{\partial q_i} = \frac{W_2 (\mathbf{q} + \delta q_i) - W_2 (\mathbf{q})}{\delta}. \qquad (9.20)$$

and we project the gradient onto the null space of the primary task, so that:

$$\dot{\mathbf{q}} = \mathbf{J}^+\dot{\mathbf{e}} + \mathbf{P_e} \, \nabla_{\mathbf{q}} W_2, \qquad (9.21)$$

which ensures that the resolution of the second task will be obtained as long as it does not interfere with the primary task.

9.2.5.2 Control of the interaction with the environment

The robot's redundancy can also be used to handle and resolve undesirable interactions with the environment. A control feature optimizes the overall configuration based on input from tactile sensors mounted along the arm or through distributed force/torque sensors. Obstacle avoidance is then activated when the sensed interaction with surrounding soft tissue reaches a defined threshold – to limit the magnitude of the physical interaction with sensitive organs, for instance.

In order to conduct experiments for such a potential extension, we equipped the virtual STIFF-FLOP arm model with distributed tactile sensors as illustrated in Figure 9.8.

When contact is detected, the strategy we propose here is to request a motion of the module in the opposite direction with respect to the central line. The direction of this motion depends thus on the sensors which have been activated (as illustrated in Figure 9.9). Note that such repulsion strategy

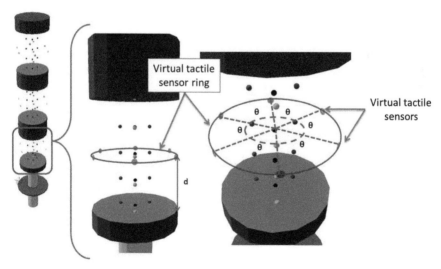

Figure 9.8 Virtual tactile sensors placed along the STIFF-FLOP arm. The tactile sensors are emulated at the locations depicted in purple (middle figure). The right figure shows sensors being equally distributed along the outer layer of the flexible structure, operating like a tactile sensor ring.

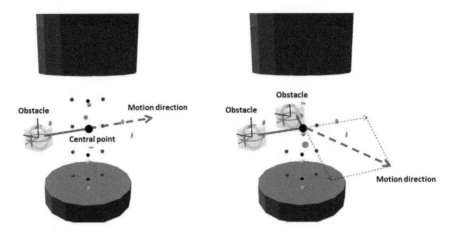

Figure 9.9   Motion request generated after contact detection by a single (left) or by multiple sensors (right).

is a proposition, and could be changed with any other function. We could, for example, accept contacts if the contact force goes beyond a predefined threshold. Any reaction strategy can be considered as long as it can be formulated as a function.

The design of a secondary task related to obstacle avoidance works as follows: Let us consider the positioning task as the priority task, and the obstacle avoidance as a secondary task. This can be formulated as:

$$\dot{q} = J^+\dot{e}+(J_o\,P_e)^+\left(\dot{q}_o-J_o\left(J^+\dot{e}\right)\right) , \qquad (9.22)$$

where $\{\dot{q}_o, J_o\}$ refers to the secondary task obstacle avoidance. The first component is set as the desired velocity to move away from the contact (as illustrated in Figure 9.9), and $J_o = \frac{\partial X_o}{\partial q}$ is the Jacobian relating the motion of the joint variables to the motion of the central point of the tactile ring. In the global equation previously described, the component $J_o\left(J^+\dot{e}\right)$ permits taking into account the motion induced by the primary task that is likely to interfere with the secondary task when combined. Therefore, since the motion direction is defined by the sensor sensing the contact (as explained in Figure 9.9), the insertion of the task for obstacle avoidance only requires defining an appropriate magnitude for the motion generated, and to be able to compute the Jacobian all along the virtual central length (assuming the tactile ring can be placed anywhere along each STIFF-FLOP module).

9.3 Inverse Kinematic Experimentations

This section demonstrates with simulations the inverse kinematics frameworks previously defined. In all experiments, the inverse kinematics model is requested to produce a bigger motion than what should be effectively requested with the real system. When controlling the real system, the requested motion is always a small displacement (in position and/or orientation) with respect to the current pose, and therefore the underlying motion request is likely to be quite small between successive motion commands. Nevertheless, for the purpose of this chapter, it is more convenient to show the inverse kinematics behavior considering larger motions.

9.3.1 Fixed Base, Various Module Representation

In these first experiments, we consider that the base is not active, i.e., the base of the first module remains fixed, and only the flexible modules are controlled to reach the target pose. The STIFF-FLOP structure considered is composed of three flexible modules.

The first illustrated experiment is a pure translational motion along the *x*- and *y*-axes. Figure 9.10 presents the initial configuration of the arm (left), and the desired final pose (right). In order to compute the appropriate

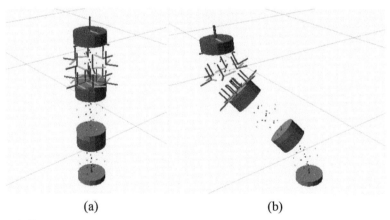

(a)   (b)

Figure 9.10 Experiment 1: Motion experiments with three modules and a fixed base, keeping a fixed orientation of the tip. (a) Initial configuration of the system. (b) Desired final configuration. The pressurized chambers are visualized through small dots between the rigid sections (gray: bottom rigid section of a module, purple: upper rigid section of a flexible module).

configuration, the information provided to the inverse kinematics is the pose of the tip frame as observed in the right figure. In the figures, the frame placed at the middle of the top pink disc corresponds to the tip location of the arm. The desired motion is a pure translation, being $[0.1, -0.1, 0.25]^\top$ with respect to the reference world frame.

In the first experimentation presented in Figure 9.11, the configuration of each module is encoded in $\mathbf{q}$ with the coordinates of each module tip position. Figure 9.11 presents details of the related minimization process by showing the evolution of the pose (position and orientation) error in the upper row, and the evolution of the tip pose along the iteration in the lower row. In all experiments (unless stated otherwise), the convergence is set to get an affine evolution of the error along the iterative minimization process, which is clearly observed with respect to the position, in the two left figures. To do so, the initial pose error is computed and, at each iteration, the algorithm is requested to compensate for a part of this error. In this experiment, the system was requested to converge within 100 iterations. Note that the completion of the requested motion can only be obtained within the pre-specified number of iterations if the Jacobian is well-conditioned.

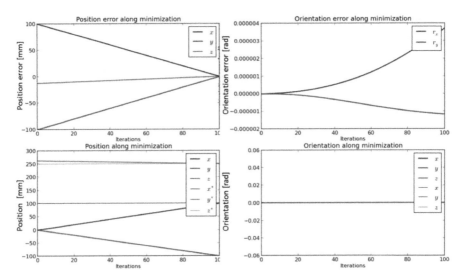

Figure 9.11   Inverse kinematic output: Upper graphs show the tip position and orientation error along the minimization process. Lower graphs present the tip position and orientation during the minimization. In the bottom graphs, the bold lines show the current values at each iteration, and the horizontal light lines show the desired values (only observable here on the position graph since the initial tip orientation is to be maintained in this example).

The evolution of the orientation error does not seem to respect the desired affine decrease along time. Looking at the magnitude of the vertical axis, we can see that the error depicted is negligible (4×10^{-6} rad, or around 0.2 millidegree), and hence can be considered to be null.

While in the experiment of Figure 9.11, the configuration of each flexible module was encoded with the module tip position, the feature used in the experiment presented in Figure 9.12 is the length of each chamber. As it can be observed, despite the use of a different feature, the minimization process is not affected, and the same pose can be obtained as well. Once more, the orientation error is sufficiently small as to be neglected.

The next experiment requires mainly an orientation adjustment of the instrument tip, which can only be obtained by bending the modules. Figure 9.13 presents the initial and final configuration considered. The targeted motion involves a rotation of around 0.6 rad (34°) around the x-axis and 0.25 rad (14°) along the y-axis. In Figures 9.14 and 9.15, we can see that both module configuration encoding methods (tip position and chamber length respectively) allow performing the desired motion. In both cases, we can see some variation of the error in position, but the observed error values are quite small (maximum of 0.04 mm and 0.20 mm) and can be neglected.

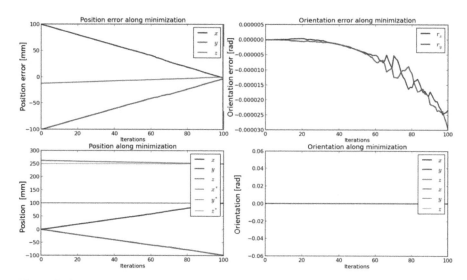

Figure 9.12 Performance of task shown in Figure 9.10 using the chamber length as model of a module configuration.

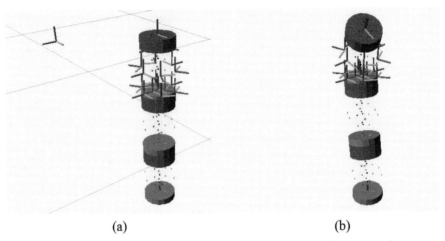

(a)   (b)

Figure 9.13   Experiment 2: Motion experiment with three modules and a fixed base. (a) Initial position. (b) Target pose, requiring only adjustment of the tip orientation.

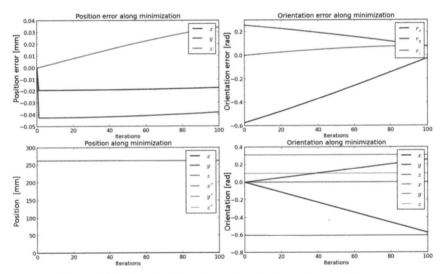

Figure 9.14   Results of Experiment 2 using tip position as a feature.

Nevertheless, we can also observe that the complete convergence towards the desired orientation is not obtained. In the tip position case (Figure 9.14), an error of around 0.1 rad (5.7°) remains for the rotation around y-axis. A similar error along the x-axis is also observed for the case concerning the chamber length (Figure 9.15). On the one hand, we can consider that the

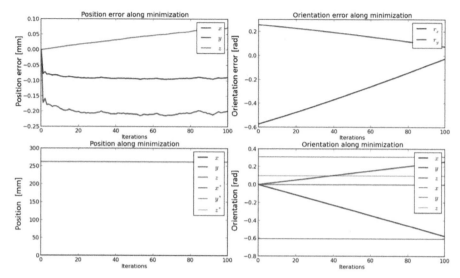

Figure 9.15 Results of Experiment 2 using chamber length as a feature.

remaining error is quite small. On the other hand, such an experiment shows the limitations of the affine convergence model. Indeed, the error value is kept constant during the whole minimization process, and a perfect convergence can only be obtained if the Jacobian is well-conditioned. Perfect convergence may thus not be observed in the bounded number of iterations. Such an issue will be commented on again when discussing the next set of experiments.

Note, finally, that in the previous experiments, where the base of the structure is fixed, we do not control the error of orientation around the z-axis, since the modules cannot directly compensate such a motion. Therefore the error function is of dimension 5. All six dimensions are considered when the arm is mounted onto a moveable base, as presented in the experiments described in the next sections.

9.3.2 Inverse Kinematics Involving the Base under Single Point Insertion Constraint

In the following experiments, we consider that the flexible modules are fixed to a rigid component which itself is connected to a robotic arm, extending the motion capabilities of the whole system. In Figure 9.16, and in the following experiments, the green component represents the connecting component that would link the STIFF-FLOP arm to a standard robotic arm. In that case,

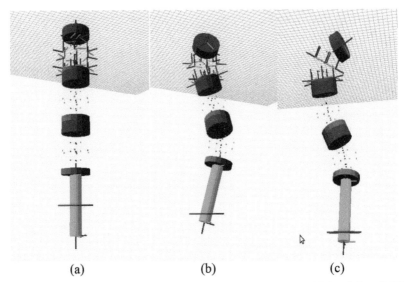

(a)   (b)   (c)

Figure 9.16 Experiment 3: Inverse kinematics considering an additional four DOF by introducing a moveable base, through motion of the green rigid rod, under the insertion point constraint. (a) Initial position (b, c) two views of the desired configuration.

the end effector of the arm would be connected to the lower section of the green cylinder (as shown in the bottom of the figure). As previously described in section 9.1.2, the system is envisioned to enter the human body through an insertion point, and the produced motion should be so that the arm always respects the virtual pivot point related to the insertion frame. In Figure 9.16, the horizontal red line represents that critical location, being the image projection of the red disc presented previously in Figure 9.6.

In the following experiments, the inverse kinematics will deduce the appropriate configuration of the flexible modules as well as the appropriate pose of the green cylinder, expressed as an extended spherical coordinate of its tip section (connected to the modules) with respect to the insertion frame. We assume then that the robotic arm will produce the appropriate motion to move the whole structure according to the computed desired pose. Note that in the following experiments the flexible module configurations are modeled with their local tip position; i.e., the joint variable q contains for each module its tip position Q_i expressed at its base.

The first experiment presented in Figure 9.16 is mainly a rotation motion. It is quite large as we can see in Figure 9.17: 0.9 rad (51°) around the x-axis and 0.2 rad (11°) around the y- and z-axes. Once more, the motion in position

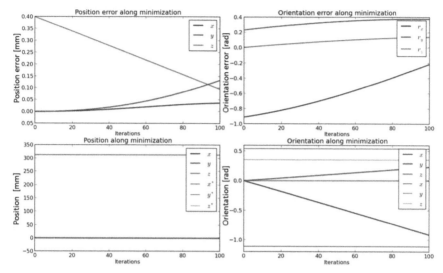

Figure 9.17   Convergence details for Experiment 3 with the moving base and the flexible module configuration encoded with the tip position.

is negligible during the minimization process (maximum of 0.40 mm). The orientation globally converges towards the desired values; nevertheless, the desired orientation is not exactly reached within the specified number of iterations. We can indeed observe a remaining error of around 0.4 rad (23°) around the x-axis and 0.2 (11°) around the y- and z-axes.

The non-completion of the task in the given timeframe is even more visible in the following experiment illustrated by Figures 9.18 and 9.19, in which the targeted displacement involves a large motion along the z-axis and a rotation around that same axis. In that case, the requested rotation is a bit more than 1.9 rad (109°).

If the error in position gets minimized on time, the error in rotation is not compensated on time. Once more the affine model shows its limit. Note that, when controlling both the base and module motions, different weights are applied to the different components, to give higher weight to the base. This indirectly affects the error minimized at each iteration and results in making the convergence impossible within the fixed number of iterations.

The most appropriate way to handle such an issue is to switch from the affine error model to an exponential model, in which the error observed is updated at each step and in which the number of iterations to convergence is not fixed. This is what has been introduced in the next experiments.

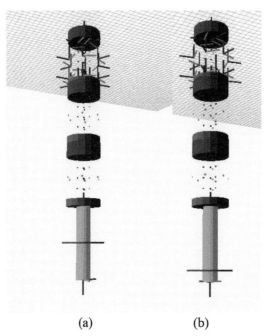

(a)   (b)

Figure 9.18   Experiment 4: Motion task involving translation along the z-axis and rotation along the z-axis. (a) Initial position. (b) Target position.

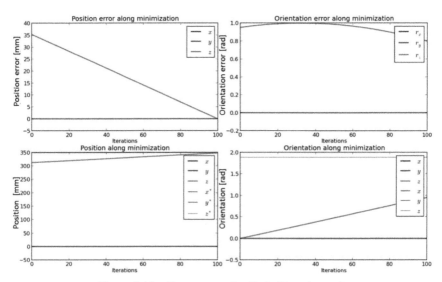

Figure 9.19   Convergence details for Experiment 4.

As mentioned in section 9.2.1, different models of error evolution can be used as part of the minimization process. All previous experiments were done with an affine error decrease ($\dot{e} = -\lambda$). Such an error model explains the evolution of the error along a straight line as observed in the previous convergence details presented. The following experiments are done considering an exponential decrease of the error ($\dot{e} = -\lambda e$).

Figures 9.20 and 9.21, respectively, show the evolution of the error for the two experiments with the active base described earlier (Experiments 3 and 4) – however, this time we consider an exponential decrease of the error. In this case, the minimization process continues until the error reaches a given precision threshold or until a maximum number of iteration is reached. In both cases, we can observe the standard exponential evolution of the error. Furthermore, the system is now able to converge more precisely to the desired poses. In such a framework, particular care has to be taken when tuning the gain parameter. A high value permits quicker convergence, but may produce an oscillation around the desired pose. A small value may reduce or even avoid the potential oscillations, but may require a much larger number of iterations to converge.

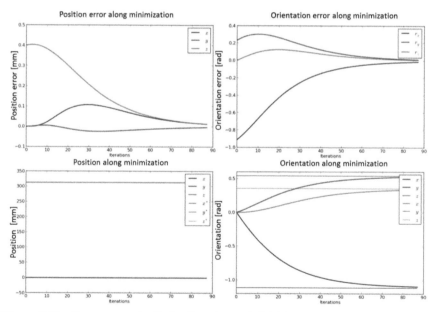

Figure 9.20 Convergence details for Experiment 3 (initial and desired poses presented in Figure 9.16), exponential evolution of the error.

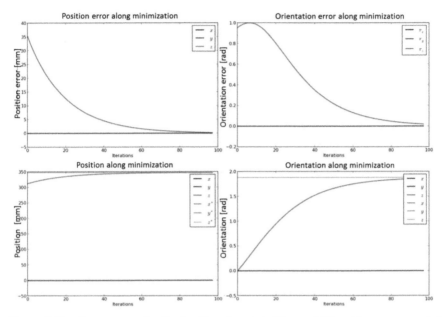

Figure 9.21   Convergence details for Experiment 4 (Figure 9.18) with an exponential evolution of the error.

9.3.3 Illustration of the Secondary Tasks

The following set of experiments illustrates the potential use of the system redundancy to control other constraints in the null space of the main task, which is the positioning one as described in section i9.2.5. The secondary tasks considered in the next experiments attempt to maintain the chambers at their mean length. Figure 9.22 presents the initial configuration and three final configurations after convergence towards the same target tip pose, but with different minimization settings for the secondary task. The three cases considered are the following ones:

- Case 1: No secondary task used,
- Case 2: Maximum elongation (α) set to 1,
- Case 3: Maximum elongation (α) set to 0.8.

The first case (1) has already been presented, and the convergence details can be seen in Figure 9.20. For completion, the convergence data are presented in Figures 9.23 and 9.24 – it is noted that the overall behavior is very similar. As expected, the secondary task is applied in the space that does not affect the first one and is related to the control of the tip of the global structure.

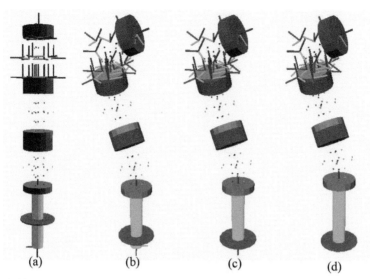

Figure 9.22 Experiment 4: Length control experimentation: (a) initial configuration, (b) obtained configuration after minimization without length control, (c) obtained configuration with first parameterization ($\alpha = 1$), (d) obtained configuration with second parameterization ($\alpha = 0.8$).

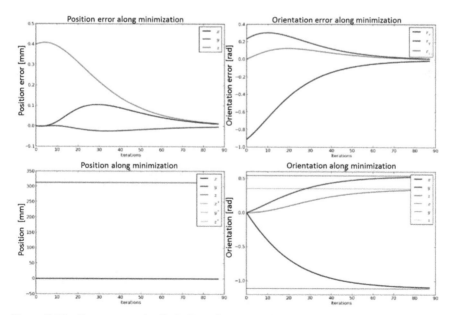

Figure 9.23 Convergence details for Experiment 4 with length control, maximum elongation set to $\alpha = 0.8$.

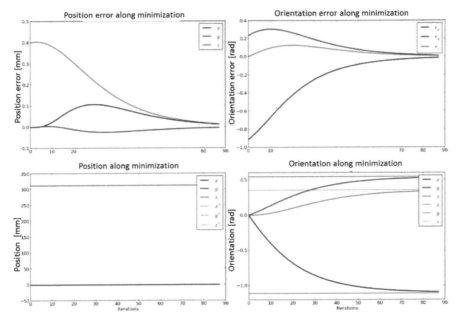

Figure 9.24   Convergence details for Experiment 4 with length control, maximum elongation set to $\alpha=1.0$.

Nevertheless, the final configurations of each module are slightly different, as can be seen from Table 9.1. The introduction of the secondary task to control the module length allows us to maintain the length of the modules to their mean dimensions as much as possible. Note that the proposed task function is designed with respect to the mean length of each module, and not directly with regard to the length of each chamber individually.

Since the modules extend their respective chambers less with regard to the overall elongation or bending of the robot, the base contributes more to the overall robot motion. This can be observed in Figure 9.22, by looking carefully at the location of the bottom extremity of the green rigid rod with respect to the red insertion disc. We can see that the base component is moving up more when the module length is being controlled. The reduced extension of the chamber length is being compensated by a larger motion of the base.

The next experiment illustrates the use of the secondary task for obstacle avoidance. The obstacle is avoided as soon as it is detected. The main

Table 9.1 Comparison of the chamber lengths obtained using the secondary task (the length at rest is 0.04 m)

Experiment	Component	Module 1	Module 2	Module 3
Case 1 (no secondary task)	Chamber 1	0.067 m	0.057 m	0.059 m
	Chamber 2	0.065 m	0.065 m	0.055 m
	Chamber 3	0.059 m	0.076 m	0.076 m
	Mean extension	58%	58.33%	58%
Case 2 $\alpha = 1$	Chamber 1	0.06 m	0.049 m	0.052 m
	Chamber 2	0.057 m	0.058 m	0.047 m
	Chamber 3	0.051 m	0.061 m	0.069 m
	Mean extension	39.33%	39.66%	39.33%
Case 3 $\alpha = 0.8$	Chamber 1	0.064 m	0.053 m	0.056 m
	Chamber 2	0.061 m	0.062 m	0.052 m
	Chamber 3	0.055 m	0.065 m	0.073 m
	Mean extension	49.33%	49.33%	49.66%

difference to the previous experiments is that the inverse kinematics is now computed in a closed-loop fashion: at each iteration, a motion request is sent to the module controllers, and the updated configuration is fed in to compute the next iteration of the minimization.

The experimental setting is similar to that of Figures 9.16 and 9.17. The convergence details obtained for this closed-loop mode, when no obstacle is detected, are presented in Figure 9.25. We can see that the trajectories are less smooth. This is mainly due to closed-loop implementation, since the dynamics of the controllers affect the minimization process (and the simulator may not be tuned sufficiently well to get a sufficiently reactive system). Nevertheless, the disturbances generated are quite small (magnitude of approximately 1 mm) and are mainly due to the fact that the emulation of the pressure variation as a function of time in the simulator may not be sufficiently well-tuned.

Figure 9.26 illustrates the settings of a similar experiment: an obstacle has been virtually placed above a tactile sensor and the system is required to move along the vertical z-axis. Figure 9.27 presents the behavior of the inverse kinematics in this case. The system moves freely along z until the tactile sensor detects the obstacle (iteration 10). The motion variation from iteration 10 to iteration 37 is related to the contribution of the obstacle avoidance task that pushes the robot modules away from the obstacle. Once avoided, the system is still able to converge towards the desired pose of the distal robot element, as we can see at the right snapshot of Figure 9.26.

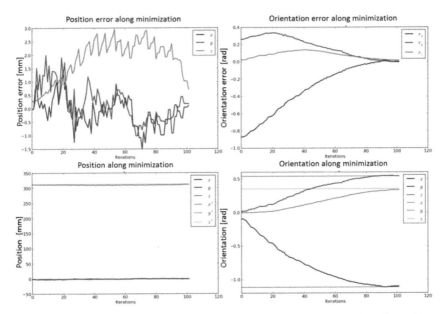

Figure 9.25 Convergence details using a closed-loop mechanism. No obstacle avoidance activated.

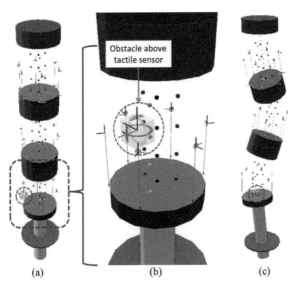

Figure 9.26 Target reaching with obstacle avoidance activated. (a) Initial configuration. An obstacle is placed just above a tactile sensor. (b) Zoom in on to the first module, the obstacle is represented through the cube. (c) Final configuration after convergence.

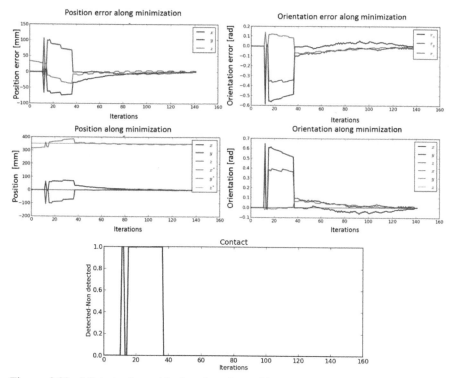

Figure 9.27 Minimization with the obstacle avoidance constraint. The lower graphic presents the activation of the secondary task, when the contact is detected.

9.4 Conclusion

This section has presented how the flexible modules can be automatically controlled to reach a target tip pose provided by the surgeon operator through a regular haptic device. Contrary to the standard configuration space using the constant curvature model, we demonstrate that the generic inverse kinematic framework proposed, based on a numerical estimation of the robot Jacobian, is able to consider other configuration spaces, such as the module tip pose or directly the chamber length. Such a framework could even be used for taking into account flexible module models that do not assume the constant curvature preservation, which seems necessary for taking into account external forces acting on the modules.

We have also shown how the inclusion of an additional standard robot could be used to extend the work space, and how it can be easily added to the inverse kinematics model. Assuming that such a robot provides position control of its end effector, we propose to embed it within the inverse kinematics framework using extended spherical coordinates that *per se* are suitable for representing the motion constraint due to the trocar location. We also mention that the traditional spherical model can be adjusted to be better shaped for surgical application, and without adding any complexity to the minimization process.

Finally, we illustrate with two specific tasks that the redundancy of the system can be used for applying, like with standard rigid-link robots, additional tasks to improve the behavior of the robot, while keeping the priority on the positioning task, that is the main request of the surgeon operator.

References

[1] Waldron, K., and Schmiedeler, J. (2008). "Kinematics," in *Springer Handbook of Robotics* (Berlin: Springer), 9–33.

[2] Webster, R. J., and Jones, B. A. (2010). Design and kinematic modeling of constant curvature continuum robots: a review. *International Journal of Robotic Research*, 29, 1661–1683.

[3] Calinon, S., Bruno, D., Malekzadeh, M. S., Nanayakkara, T., and Caldwell, D. G. (2014). Human-robot skills transfer interfaces for a flexible surgical robot. *Computer Methods and Programs in Biomedicine*, 116, 81–96.

[4] Guthart, G. S., and Salisbury, K. J. (2000). "The intuitive telesurgery system: overview and application," in *Proceedings of the IEEE International Conference on Robotics Automation*, Vol. 1, 618–621.

[5] Zemiti, N., Morel, G., Ortmaier, T., and Bonnet, N. (2007). "Mechatronic design of a new robot for force control in minimally invasive surgery," in *Proceedings of the IEEE/ASME Transactions on Mechatronics*, 12, 143–153.

[6] Unger, S. W., Unger, H. M., and Bass, R. T. (1994). AESOP robotic arm. *Surgical Endoscopy*, 8:1131.

[7] Ghodoussi, M., Butner, S. E., and Wang, Y. (2002). "Robotic surgery – The transatlantic case," in *Proceedings of the IEEE International Conference on Robotics and Automation*, 2, 1882–1888.

[8] Aghakhani, N., Geravand, M., Shahriari, N., Vendittelli, M., and Oriolo, G. (2013). "Task control with remote center of motion constraint for minimally invasive robotic surgery," in *Proceedings of the IEEE International Conference on Robotics and Automation*, 5807–5812.

[9] Nasseri, M. A., Gschirr, P., Eder, M., Nair, S., Kobuch, K., Maier, M., et al. (2014). "Virtual fixture control of a hybrid parallel-serial robot for assisting ophthalmic surgery: an experimental study," in *Proceedings of the 5th IEEE RAS and EMBS International Conference on Biomedical Robotics and Biomechatronics*, 732–738.

[10] Ma, J., and Berkelman, P. (2006). "Control software design of a compact laparoscopic surgical robot system," in *Proceedings of the 2006 IEEE/RSJ International Conference Intell Robot Systems*, 2345–2350.

[11] Osa, T., Staub, C., and Knoll, A. (2010). "Framework of automatic robot surgery system using visual servoing," in *Proceedings of the IEEE/RSJ International Conference on Intelligent Robots and Systems (IROS)*, 1837–1842.

[12] Staub, C., Panin, G., Knoll, A., and Bauernschmitt, R. (2010). Visual instrument guidance in minimally invasive robot surgery. *International Journal on Advances in Life Sciences*, 2, 103–114.

[13] Muñoz, V. F., Gómez-de-Gabriel, J. M., García-Morales, I., Fernández-Lozano, J., and Morales, J. (2005). Pivoting motion control for a laparoscopic assistant robot and human clinical trials. *Advanced Robotics*, 19, 694–712.

[14] Uecker, D. R., Lee, C., Wang, Y. F., and Wang, Y. (1995). Automated instrument tracking in robotically assisted laparoscopic surgery. *Journal of Image Guided Surgery*, 1, 308–325.

[15] Chen, G. (2005). *Design, Modeling and Control of a Micro-robotic Tip for Colonoscopy.* Ph.D., thesis, Institut National des Sciences Appliquées, Lyon.

[16] Fraś, J., Czarnowski, J., Maciaś, M., and Główka, J. (2014). "Static modeling of multisection soft continuum manipulator for stiff-flop project," in *Proceedings of the Advances in Intelligent Systems and Computing*, (Cham: Springer), 365–375.

[17] Remazeilles, A., Prada, M., and Rasines, I. (2017). "Appropriate spherical coordinate model for trocar port constraint in robotic surgery," in *Proceedings of the Iberian Robotics Conference*.

10

Modelling and Position Control of the Soft Manipulator

Jan Fras, Mateusz Macias, Jan Czarnowski and Jakub Glowka

Industrial Research Institute for Automation and Measurements PIAP, Warsaw, Poland

Abstract

This chapter presents the numerical kinematics model of the manipulator developed during the STIFF-FLOP project. The model is based on the Euler-Bernoulli beam theory and Hooke's law and provides the manipulator shape approximation based on the sensor readings and pressure input values. The described model has been numerically inverted and used for model-based position and orientation control for the manipulator.

10.1 Introduction

The STIFF-FLOP manipulator is designed to operate in tight spaces in which the appearance of external forces is highly probable. Therefore the constant curvature model is no longer applicable since its results are non-satisfactory (see Figure 10.1) in terms of manipulator shape assessment and control. The shape of the manipulator can be simulated and controlled using complex algorithms like the Finite Element Method, but methods like this are very computationally expensive and are also not applicable in our case. Because of this, a new shape calculation method has been developed [1]. The solution is based on physical principles like Hooke's Law and Euler-Bernoulli beam theory. It simulates the module shape using combined information about its elongation, bending, and twist along the module axis. The model (also called Bending Model) assumes that those parameters can vary along the module shape.

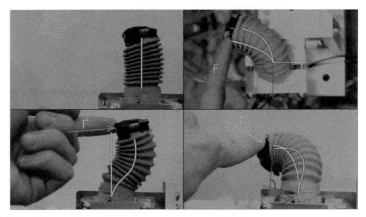

Figure 10.1 Cases where the constant curvature model is not valid. External forces make the module bend and its shape does not resemble a fragment of the circle. The blue vector represents the applied external force, the red line is an approximation of module shape using the constant curvature model, and the green line is an approximation of module shape.

10.2 Assumptions

For modeling purposes, simplifying assumptions have been made as follows:

- Pressures in each chamber and the values of external forces acting on the module are measured at any point of the module;
- Module is made of homogeneous material of known stiffness, with three pressure chambers hollowed out;
- The dimensions of the cross-section of the pressure chambers are constant (achieved through braiding);
- Pressures in chambers are constant at any point.

We have also neglected the influence of the reinforcement structure and other components that might be embedded into the module construction such as sensors, etc. Still the influence of the omitted factors exists and may introduce some errors into the shape approximation. The STIFF-FLOP manipulator is built from three identical modules. Equations and calculation are presented for a single module. The model could be applied to all segments of the manipulator allowing the final shape of the structure to be estimated.

10.3 Single Segment Model

The STIFF-FLOP manipulator is driven by pneumatic pressure delivered into its actuation chambers. As the chambers are relatively small, in the static conditions the pressure in each chamber has almost constant value at any point of its volume. The pressure inside the considered chamber can be treated as an internal stress and described by Equation (10.1), where p denotes the pressure value, F a force value acting on a part of cross-sectional area A.

$$p = \frac{F}{A} \tag{10.1}$$

An integral of such stress over the chamber cross-section gives the value of an overall force acting in the center of the chamber and perpendicular to such a cross-section. Integrating pressures from all the chambers, all the pressure-related forces acting in the cross-section can be calculated. Since in our assumptions the module geometry is constant throughout its length and the pressure is constant as well, the resulting forces have the same value in any cross-section perpendicular to the module axis. To better present the idea, the cross-section of the module is presented in Figure 10.2. As the location and direction of force values is known, the bending moment acting in the cross-section and the stretching force can be calculated. As all the forces are parallel to the manipulator central axis, the resulting bending moment z-component equals zero, which indicates that no twisting force is present. The bending moment acting in the considered cross-section k can be assessed as Equation (10.2), where C_{ki} denotes a vector from the geometrical cross-section center to the center of i-th chamber cross-section center for the k-th cross-section. F_{ki} stands for the force acting on the i-th chamber cross-section.

$$\vec{M_{kp}} = \vec{C_{k1}} \times \vec{F_{k1}} + \vec{C_{k2}} \times \vec{F_{k2}} + \vec{C_{k3}} \times \vec{F_{k3}} \tag{10.2}$$

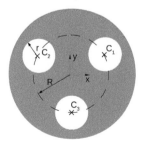

Figure 10.2 Cross-section of the module.

Deformation related to the bending moment Equation (10.2) can be expressed using Euler-Bernoulli formula Equation (10.3).

$$\kappa = \frac{1}{\rho} = \frac{M}{EI} \tag{10.3}$$

where E is the Young's modulus constant of the silicone, and I is the second moment of inertia of the module at cross section k.

If no external force is acting on the module, the bending moment has the same value at every cross section. In such case the curvature value is equal for each cross-section Equation (10.4).

$$\forall k, \kappa_k = \kappa_0, \kappa_0 = \frac{M_0}{EI} \tag{10.4}$$

where $\frac{\kappa_k=1}{\rho_k}$ stands for the curvature at the cross-section k.

Such situation is presented in Figure 10.3 and can be described by the constant curvature model that assumes the module axis shape to be a part of a circle [2].

Forces resulting from internal applied pressures also influence the module's length. Elongation at any point along the module's axis can be described using the Hooke's Law Equation (10.5).

$$\Delta dl = \frac{F_p}{AE} dl \tag{10.5}$$

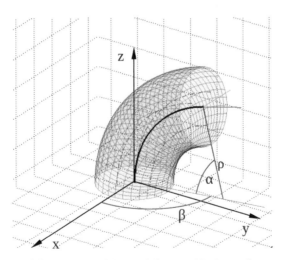

Figure 10.3   Bending of the module caused by internal pressures.

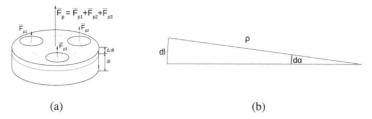

(a)   (b)

Figure 10.4 Elongation and bending of small section of the manipulator.

where F_p stands for the overall pressure-related force acting in the considered cross-section that is equal to the sum of individual forces acting in chambers. As mentioned, this force only has a z component. Any radial deformation related to the pressure is constrained by the manipulator structure and reinforcement (Chapter 3). A denotes the area of the silicone part of the cross section (cross section area minus the area of chambers). In Figure 10.4(a) a part of the manipulator elongated by pressures is presented. The overall module elongation can be calculated by integrating Equation (10.5) from 0 to l_0: Equation (10.6), where l_0 is the length of non-actuated module.

$$\Delta l = \int_o^{l_0} \Delta dl = \int_0^{l_0} \frac{F_p}{AE} dl \tag{10.6}$$

In order to calculate the overall bending angle α the integral of curvature from 0 to $l + \Delta l$ can be used Equation (10.7) – see Figure 10.4(b).

$$\alpha = \int_0^{l_0+\Delta l} d\alpha = \int_0^{l_0+\Delta l} \frac{1}{\rho} dl \tag{10.7}$$

10.4 External Forces

If there are any external forces acting on the manipulator, the equations given above (10.1) to (10.7) have to be modified to take those forces into account. It is important to note that the constant curvature model does not apply any longer, since the bending moment resulting from forces (internal and external) may vary in different cross-sections. The updated equation for the bending moment would be Equation (10.8).

$$\vec{M}_k = \vec{M}_{kp} + \vec{M}_{k \; ext} \tag{10.8}$$

where $M_{k\ ext}$ is the bending moment component present in k^{th} cross-section resulting from the external forces Equation (10.9).

$$\vec{M}_{k_{ext}} = \vec{F}_{ext} \times \vec{r}_k \tag{10.9}$$

where $\vec{r}_k$ is a vector form the center of the kth cross-section to the point the force F_{ext} is applied.

Considering the pressure and the external forces, impact on the manipulator, the aggregated deformation can be expressed as Equation (10.10),

$$\vec{\kappa}_k = \frac{1}{\vec{\rho}_k} = \frac{\vec{M}_k}{EI} \tag{10.10}$$

where:

$$\vec{M}_k = \vec{M}_{kp} + \vec{F}_{ext} \times \vec{r}_k$$

and overall elongation as Equation (10.11)

$$\Delta dl_k = \frac{F_k}{AE} dl \tag{10.11}$$

where:

$$F_k = F_p + \vec{F}_{k\ ext,z}$$

$F_{k\ ext,z}$ stands for the external z axis component of the external force expressed in the kth cross-section reference frame, Figure 10.5.

In the general case, apart form the bending moment and stretching force, the module could also be affected by a twisting moment coming from external

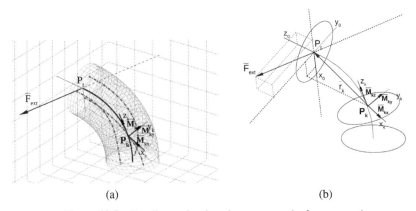

(a)   (b)

Figure 10.5   Bending and torison in respect to the k cross-section.

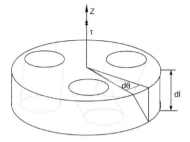

Figure 10.6 Torsion of the module fragment *dl*.

forces. Deformation resulting from such a moment can be expressed as Equation (10.12)

$$d\frac{\theta}{dl} = \frac{T}{GI_0} \tag{10.12}$$

where T stands for the torque moment, G denotes Kirchhoff module, and I_0 is the polar second moment of inertia of the module cross-section.

The elementary twist in k^{th} cross-section has been presented in Figure 10.6.

10.5 Analytical Issues

Solving the above integrals is not trivial. General mechanical methods for calculating the deflection of loaded objects assume that the deflection is small (like α_1 in Figure 10.7). Assumption of small deflection allows us to integrate over the length of the undeformed object, i.e., from l to 0 by dx, Equation (10.13). This is because for small deflections $\alpha = \alpha_1$ lengths dx and dl_1 are approximately the same Equation (10.14), however, when the deflection gets bigger ($\alpha = \alpha_2$) such a relationship does not apply [Equation (10.15)]. Thus, when the deflection is not small such methods do not perform well. The assumption of small deflection is obviously not fulfilled in case of the soft STIFF-FLOP manipulator and for that reason such methods can not be easily applied.

Thus in our case the manipulator module is divided into an arbitrary number of cross-sections and the overall deformation can be efficiently modeled by integrating deformations in each of the cross-sections.

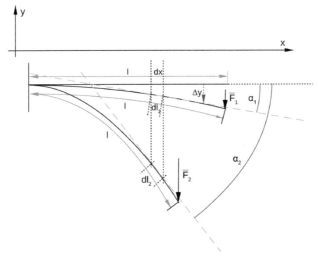

Figure 10.7   Integration of deflection, standard strength of materials approach. The assumption of small deflection is required, so that integration by dl equals to integration by dx and $dl||dx$.

$$y(x) = -\int \left[\int \frac{M_g(x)dx}{EI_z} \right] dx + Cx + D \qquad (10.13)$$

where C and D are constants

$$dl_1 = \frac{dx}{\cos(\alpha_1)} \approx \frac{dx}{1} = dx, \; \alpha_1 \approx 0 \qquad (10.14)$$

$$dl_2 = \frac{dx}{\cos(\alpha_2)} > \frac{dx}{1} = dx, \; |\alpha_2| \gg 0 \qquad (10.15)$$

The described model has been implemented in MATLAB and Python. The calculations are done in the direction from the module's tip to its base. In each iteration, bending moments and forces acting on the particular cross-section are calculated in the cross-section reference frame. Based on those values, the elongation, bending, and twist of the corresponding module slice is calculated as Equations (10.16) and (10.18).

$$\Delta l = \frac{L}{N} \left(1 + \frac{F_{nz}}{AE} \right) \qquad (10.16)$$

$$\Delta\theta = \Delta l \frac{M_{nxy}}{EI_{xy}} \qquad (10.17)$$

$$\Delta\beta = \Delta l \frac{M_{nz}}{GI_0} \tag{10.18}$$

where:

- Δl - length of the considered manipulator slice
- $\Delta\beta$ - deflection corresponding to the slice,
- $\Delta\beta$ - twist corresponding to the slice,
- N - arbitrary number of slices – the more the more precise the result,
- L - rest length of the module,
- F_{nz} - z component of the force acting on n^{th} cross-section,
- A - cross-section area, E – Young's modulus,
- M_{nxy} - length of bending moment vector projected on XY plane in the cross-section reference frame,
- I_{xy} - second moment of inertia of the module cross-section,
- M_{nz} - z component of the bending moment in the cross-section reference frame,
- G - Kirchhoff module,
- I_0 - polar second moment of inertia of the module cross-section.

The direction of the bending can be asessed using Equation (10.19).

$$\theta = a \tan 2(M_y, M_z) \tag{10.19}$$

Using the above equations the transformation from an nth cross-section to cross-section $(n+1)^{\text{th}}$ can be obtained.

10.6 Inverse Kinematics

Since the constant curvature assumption is not valid in case of any external forces acting on the manipulator (Figure 10.1), the inverse kinematics based on such assumption also can not provide accurate results. We would like to introduce another approach based on the Bending Model, which has been presented earlier in Section 10.3. Since the model is numerical, it is not simple to find its inverse function. The solution we propose is to use numerical optimization algorithms in order to find module configuration which fulfils the required tip position and orientation. The algorithm takes the position and the values of external forces as input, and calculates pressures which guarantee reaching the goal position in these specific conditions. The calculations can be performed for any number of modules. The principle of the algorithm is minimization of the cost function which is the tip-to-goal distance [Equation (10.20)]. The minimization is carried out in several steps, in which a different set of pressures is tested. The algorithm stops when the tested set of pressures guarantees that the achieved tip position is close enough to the desired goal.

The most important issue in such an approach is the method of deduction of the pressure values to be tested in the upcoming step. The efficiency of the whole algorithm depends on the number of iterations required to be executed to find a satisfying solution. The proposed method for minimization of the tip position error is adjusting the current pressure values based on gradient descent algorithm.

In general, the method starts at some point in the pressure space. Next, the pressure values for the next step are calculated using the current gradient value [Equations (10.21)] and 10.22]. For the calculation, the gradient of the manipulator tip position error is estimated for relatively small pressure changes.

$$f(x) = dist(M(p_1, p_2...p_3, f_1, f_2...f_3), x_0, y_0, z_0),$$
$$\text{where} : x = [p_1...p_n, f_1...f_n, x_0, y_0, z_0] \tag{10.20}$$

$$p_{i+1} = p_i + \Delta p = p_i + [\Delta p_1 \Delta p_2 \cdots \Delta p_n] = p_i + \frac{f(x)}{|\nabla(f(x))|} \cdot \vec{\nabla}(f(x)) \tag{10.21}$$

$$\vec{\nabla}(f(x)) = \left[\frac{df(x)}{dx_1} \frac{df(x)}{dx_2} \cdots \frac{df(x)}{dx_n} \right]^T \tag{10.22}$$

The distance function (which is the cost function in the described case) is not linear in terms of the value of the pressure change. Therefore, in most cases several steps have to be evaluated to find the solution. Moreover, in some areas the pressures calculated using [Equation (10.21)], changes the tip position too much, which can be the reason for the algorithm instability.

Based on the experimental results, the additional factor a has been introduced [Equation (10.23)]. Keeping this factor value between 0 and 1 makes the convergence slower at the beginning, but more stable in close goal areas, and finally reduces the required number of steps.

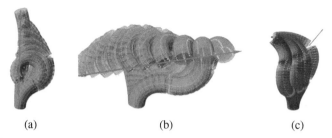

      (a)   (b)   (c)

Figure 10.8 Inverse Kinematics algorithm visualization. The blue arrow denotes the path followed by the tip of the manipulator, the red one denotes a force acting on the manipulator's body.

$$p_{i+1} = p_i + \Delta p \cdot a,$$
$$\text{where } 0 < a < 1 \tag{10.23}$$

The results of the inverse kinematics applied to a manipulator build from three identical modules are shown in the Figure 10.8. In Figures 10.8(a) and 10.8(b) the manipulator's tip linear movement is presented and in the Figure 10.8(c) the manipulator's response to the changing external force value which is acting on its surface.

Since there are many solutions to this issue, the algorithm finds the nearest one in the solution space. There is a possibility that the algorithm will not find the solution even if one exists. That is because local minima can occur.

10.7 Conclusion

The modeling strategy presented in this chapter has been successfully used for reconstruction of the STIFF-FLOP spatial configuration. The proposed solution provides better results than the basic strength of materials methods and is far less computationally complex than methods based on finite elements. It takes both the internal and external factors into account and calculates deformation of the STIFF-FLOP robot that includes elongation, bending, and twist. We were able to successfully utilize the proposed model for real-time inverse kinematics of the manipulator embedded into the robot controller. The final structure of the manipulator used in the project was a bit different from the one assumed in this chapter (see Chapter 3); however, adjusting the model was as simple as adjusting moment of inertia values for corresponding cross-sections. The developed modeling and control strategy can be easily adapted to any continuum manipulator.

Acknowledgments

The work described in this chapter is supported by the STIFF-FLOP project grant from the European Commission's Seventh Framework Programme under grant agreement 287728. This project is also partly supported from funds for science in the years 2012–2015 allocated to an international project cofinanced by Ministry of Science and Higher Education of Poland.

References

[1] Fraś, J., Czarnowski, J., Maciaś, M., and Główka, J. (2014). "Static modeling of multisection soft continuum manipulator for stiff-flop project," in *Proceedings of the Recent Advances in Automation, Robotics and Measuring Techniques* (Berlin, Heidelberg: Springer), 365–375.

[2] Webster, III, R. J., and Jones, B. A. (2010). Design and kinematic modeling of constant curvature continuum robots: a review. *Int. J. Rob. Res.* 29, 1661–1683.

11

Reactive Navigation for Continuum Manipulator in Unknown Environments

Ahmad Ataka[1], Ali Shiva[1] and Kaspar Althoefer[2]

[1]King's College London, London, United Kingdom
[2]Queen Mary University of London, London, United Kingdom

Abstract

Navigating continuum robots, such as STIFF-FLOP manipulator [7], to reach a certain configuration without colliding with its surrounding environments tends to be an ongoing research area as the inherent complexity of this class of robots poses new challenges in this field compared to rigid-link robots. In this chapter, we describe the reactive navigation algorithm for guiding the tip of continuum robot toward the goal while, at the same time, avoiding the robot body from collision with the surrounding environment. We limit the navigation problem in this chapter to the case where the information of the environment is unknown before movement execution and the robot can rely only on the most updated sensory information. Two navigation algorithms, both inspired by the physical phenomena of electromagnetism, are implemented in the kinematic model of multisegment continuum manipulator. The pose estimation strategy needed to estimate the position of the manipulator body is also presented.

11.1 Introduction

Inspiration from certain species in nature such as the octopus [1] has sparked a new trend in robotics which aims at enhancing robotic maneuverability and dexterity compared to the conventional rigid-link robots, giving birth to the new breed of continuum robots. Many continuum robotic systems have been developed during the recent years [2–4] using a variety of designs both for the structure and for the actuation methods, such as tendon driven systems [5], pneumatically actuated robots [6], or stiffness-controllable manipulators [4]. A recent example of this new type of robot has been presented in [7], where the STIFF-FLOP manipulator, built upon soft material, is designed to be used for a surgical purpose. Although developments in this field show potential improvement in the features which aim to resemble biology, yet the structural flexibility of this new class of robots poses new challenges in areas such as modeling, control, pose estimation, and navigation.

The subject of robot navigation continues to be the focus of many research works. Yet, with regard to continuum manipulators, research in this area is still in its early stages and therefore not fully conclusive. A common drawback witnessed in a majority of works based on optimization-based planning [1, 3, 4, 6, 8] and sampling-based planning [9–13] is the lack of capability in dealing with unknown obstacles. In [14], an inverse kinematic method is described for a steerable needle. However, due to the assumption of a static and well-defined environment, this method falls short of dealing with unknown obstacles. Some other studies which aim to tackle dynamic and unknown environments, although provide interesting solutions, are yet constrained to specific geometries and/or applications. For instance, an adaptive motion planning algorithm is discussed in [12] for the OctArm manipulator, but is limited to a planar scenario. The inverse-Jacobian method proposed in [15] specifically addresses guidance inside a tubular environment. In [16], the null space of a redundant continuum manipulator is exploited for navigating the robot. However, the method is not computationally cost efficient due to successive matrix multiplications.

One of the prominent trends in robot navigation is taking inspiration from natural physical phenomena such as the electric potential field [17]. Being relatively simple and straightforward, this method has been extensively applied for mobile robots, rigid manipulators, and even continuum manipulators [18, 19]. The global convergence is not guaranteed in this algorithm. However, since this method does not rely on the global map of the environment, it can be used in unknown and dynamic environments, and hence applied in a reactive manner.

Other physics-inspired works applied the properties of an artificially induced magnetic field for navigating rigid-link manipulators, such as [20, 21]. Although this method claims to resolve the problem of entrapment in local minima observed in the electric potential field case, this method still requires the geometry and location of obstacles to be known beforehand thus being impractical for unknown and/or dynamic obstacles. The effort to make the magnetic-field-inspired approach more reactive is set forward in [22] by defining the concept of circular fields which can be applied to partially unknown environments. This method, however, requires a prior knowledge of the location of the center of the obstacles, which would not be possible to implement in real-world scenarios with unknown obstacles.

In this article, we present our recent works on continuum manipulator navigation which is inspired by the physical phenomena of electromagnetism. The use of electric field to keep manipulator body from collision is based on our works in [18]. Another method presented here is the use of magnetic field to avoid the collision, in which the backbone of the manipulator is considered as the means for current flow, resembling an electrical wire which induces a field on nearby objects. It assumes an artificial current flowing continuously along the segment of the continuum manipulator which induces "shadow" current on the surface of the nearby obstacle in such a way that the system behavior will resemble the physical behavior of the two current-carrying wires. Combined with attractive potential function at the tip and the pose estimation strategy to estimate the pose of any point along the body of the manipulator as presented in [19, 23], both methods will guide the manipulator's tip at a desired target location while at the same time prevent the manipulator body from colliding with unknown environment in a reactive manner.

11.2 Modeling and Pose Estimation

11.2.1 Kinematic Model

The constant curvature kinematic model is used in this research. The underlying assumption is that each manipulator's segment can be mathematically described by the equation of circular arc with a constant radius of curvature. Therefore, every segment can be fully characterized by configuration space variables $\mathbf{k}_i = \begin{bmatrix} \kappa_i & \phi_i & s_i \end{bmatrix}^T$, denoting the curvature (κ_i), rotational deflection angle (ϕ_i), and arc length (s_i) of segment-i, respectively, as illustrated in Figure 11.1a. The tip of segment-i with respect to the base, $^i_{i-1}\mathbf{T}(\mathbf{k}_i) \in SE(3)$, will depend solely on variable $\mathbf{k}_i$ as explained at length in [11]. In this chapter, we assume that the manipulator's base is fixed. Hence, the pose of the end effector with respect to the world frame attached to the base of the first segment for manipulator with N segments is described as

$$^N_0\mathbf{T}(\mathbf{k}) = \prod_{i=1}^{N} {}^i_{i-1}\mathbf{T}(\mathbf{k}_i), \tag{11.1}$$

where $\mathbf{k} = \begin{bmatrix} \mathbf{k}_1 & \mathbf{k}_2 & \dots & \mathbf{k}_N \end{bmatrix}^T$.

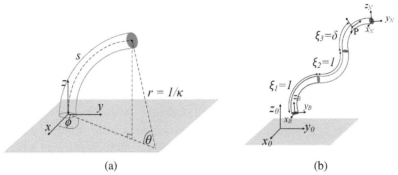

(a)   (b)

Figure 11.1 (a) The frame definition attached to each individual segment. (b) Illustration of a three -segment continuum manipulator with a movable base.

These configuration space variables **k** are normally different from the real controllable parameters of the actuator and generally very robot-specific due to the wide range of choices in design and actuating mechanism. For a tendon-driven robot like the one used in [18], the actuator space variables are the tendon length and can be written as $\mathbf{q}_i = \begin{bmatrix} l_{i1} & l_{i2} & l_{i3} \end{bmatrix}^T$ in which l_{ij} denotes the length deviation of tendon-j in segment-i from normal length L. For N-segment manipulator, the whole actuator space is expressed as $\mathbf{q} = \begin{bmatrix} \mathbf{q}_1 & \mathbf{q}_2 & \cdots & \mathbf{q}_N \end{bmatrix}^T$. The mapping between configuration space variables of segment-i, $\mathbf{k}_i$, and tendon length in segment-i, $\mathbf{q}_i$, depends on the manipulator's cross-section radius d as has been derived in [11].

To help expressing the pose of any point along the body of the manipulator, the scalar coefficient vector is employed as illustrated in Figure 11.1b. To specify a point in segment-i, a continuous scalar $\xi_i \in [0, 1]$ is used which covers all the point from the base ($\xi_i = 0$) to the tip ($\xi_i = 1$) of the segment. The set of scalars from all segments constitutes a vector, $\xi = \begin{bmatrix} \xi_1 & \xi_2 & \cdots & \xi_N \end{bmatrix}^T$ defined as $\xi = \{\xi_r = 1 : \forall r < i, \xi_i, \xi_r = 0 : \forall r > i\}$ [16].

Using the above formulation, the forward kinematic is written as

$$\,_0^N\mathbf{T}(\mathbf{q}, \xi) = \begin{bmatrix} \mathbf{R}(\mathbf{q}, \xi) & \mathbf{p}(\mathbf{q}, \xi) \\ \mathbf{0}_{1 \times 3} & 1 \end{bmatrix} \tag{11.2}$$

where $\mathbf{R}(\mathbf{q}, \xi) \in SO(3)$ and $\mathbf{p}(\mathbf{q}, \xi) \in \mathbb{R}^3$, respectively, express the orientation and position of any point along the body of the manipulator.

A Jacobian matrix, $\mathbf{J}(\mathbf{q}, \xi) \in \mathbb{R}^{3 \times (3N)} = \frac{\partial \mathbf{p}(\mathbf{q}, \xi)}{\partial \mathbf{q}}$, transforms the velocity vector as follows

$$\dot{\mathbf{p}}(\mathbf{q}, \xi) = \mathbf{J}(\mathbf{q}, \xi)\dot{\mathbf{q}} \Leftrightarrow \dot{\mathbf{q}} = \mathbf{J}(\mathbf{q}, \xi)^{-1}\dot{\mathbf{p}}(\mathbf{q}, \xi). \tag{11.3}$$

Another way of describing the kinematic model of continuum manipulator is by using the state space representation described by the following equations

$$\mathbf{x}_{k+1} = f(\mathbf{x}_k, \mathbf{u}_k), \tag{11.4}$$

$$\mathbf{y}_k = g(\mathbf{x}_k), \tag{11.5}$$

where $\mathbf{x}_k \in X$, $\mathbf{u}_k \in U$, and $\mathbf{y}_k \in Y$, respectively, denote the state variable, input, and output value at iteration-k, X, U, and Y, respectively, denote the state space, input space, and output space, while $f : X \times U \to X$ and $g : X \to Y$, respectively, map the current state and input value to the next state and the current state to the output value. Here we define the term *active segments* to denote $n \leq N$ number of segments which will be analyzed using state space representation.

To transform the kinematic model into a state space equation, the tendon's actuator space $\mathbf{q} \in \mathbb{R}^{3n}$ is chosen as state variables $\mathbf{x}$ while the tendon length's rate of change $\dot{\mathbf{q}} \in \mathbb{R}^{3n}$, proportional to the rotational speed of the motor to which the tendon is connected, is the input $\mathbf{u}$. The state and input variables are expressed as

$$\mathbf{x} = \mathbf{q} = \begin{bmatrix} \mathbf{q}_1 & \mathbf{q}_2 & \cdots & \mathbf{q}_n \end{bmatrix}^T, \tag{11.6}$$

$$\mathbf{u} = \dot{\mathbf{q}} = \begin{bmatrix} \dot{\mathbf{q}}_1 & \dot{\mathbf{q}}_2 & \cdots & \dot{\mathbf{q}}_n \end{bmatrix}^T, \tag{11.7}$$

in which $\mathbf{q}_i = \begin{bmatrix} l_{i1} & l_{i2} & l_{i3} \end{bmatrix}^T$. Therefore, using this definition, the state equation in Equation (11.4) describing the kinematic model of continuum manipulator is written as

$$\mathbf{x}_{k+1} = f(\mathbf{x}_k, \mathbf{u}_k) = \mathbf{x}_k + \Delta t \mathbf{u}_k, \tag{11.8}$$

where the time sampling is denoted by Δt.

The output variable $\mathbf{y} \in \mathbb{R}^{3n}$ in Equation (11.5) is represented by the tip position of each active segment retrieved from a 3-DOF position sensor embedded in the tip of each segment.

$$\mathbf{y}_k = g(\mathbf{x}_k) = \begin{bmatrix} \mathbf{p}(\mathbf{x}_k, \xi = \chi_1) & \cdots & \mathbf{p}(\mathbf{x}_k, \xi = \chi_n) \end{bmatrix}^T, \tag{11.9}$$

in which $\chi_i = \begin{bmatrix} \xi_1 = 1 & \cdots & \xi_i = 1 & \xi_{i+1} = 0 & \cdots & \xi_n = 0 \end{bmatrix}^T$ while $\mathbf{p}(\mathbf{x}_k, \xi)$ is taken from Equation (11.2).

11.2.2 Pose Estimation

In order to avoid the manipulator body from collision, the pose of any point along the body of the robot needs to be known at every iteration. This is not directly possible without the sensing capability of each segment's tendon length. The state space representation of the continuum robot model described in the previous subsection is beneficial to estimate the pose of any point along the body of the manipulator with only position sensor available embedded in the tip of each segment.

The well-known extended Kalman filter (EKF) is modified to estimate the tendon's length of each segment using the kinematic model and the measured position data of the tip of each segment. The state space equations expressed in Equations (11.8) and (11.9) enable us to estimate the next tendon length value $\hat{\mathbf{x}}_{k+1|k+1}$ based on the current estimate $\hat{\mathbf{x}}_{k|k}$, input $\mathbf{u}_k$, and sensor data $\mathbf{y}_k$ using the information of the estimation covariance $\mathbf{P}_{k|k}$, process noise variance $\mathbf{Q}_k$, measurement noise variance $\mathbf{R}_k$, and linearized model of state Equations (11.8) and (11.9) characterized by matrix $\mathbf{A}_k = \frac{\partial f(\mathbf{x}_k, \mathbf{u}_k)}{\partial \mathbf{x}_k}$ and $\mathbf{C}_k = \frac{\partial g(\mathbf{x}_k)}{\partial \mathbf{x}_k}$.

As explained in more detail in our previous work [23], implementing the standard EKF could make the state estimate arrives at physically impossible state values, for instance, a negative tendon length, caused by the multiple mathematically possible states for a certain segment tip position. Therefore, a multistage implementation of EKF, as illustrated in Figure 11.2, is employed. The underlying principle of this approach is by taking a segment into the estimation process only when the tip pose estimation error of the previous segment is smaller than a threshold δ. The full description of the algorithm

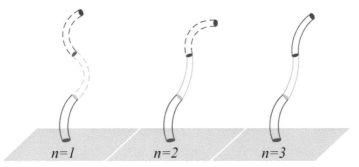

Figure 11.2   Illustration of the multistage extended Kalman filter for the case of three-segment continuum manipulator.

is shown in Algorithm 1 and described at length in [23]. The state estimate which represents the estimated tendon length $\hat{\mathbf{x}}$ produced by EKF can then be used in Equation (11.2) to determine the estimated pose of every point along the manipulator body needed by the navigation algorithm which will be explained in the next section.

11.3 Reactive Navigation

In this section, we present the reactive navigation algorithm inspired by the phenomena of electromagnetism. The underlying principle of this approach is that the manipulator moves in space filled with an artificial vector field $\mathbf{F}(\mathbf{q})$ designed to make the manipulator behave in an intended way to avoid collision with surrounding obstacle. Besides that, the end effector of the manipulator will also be influenced by an artificial attractive vector field toward the intended goal position.

Algorithm 1 Multistage pose estimation.

1:   $n \leftarrow 1, k \leftarrow 0$
2: **loop**
3:     $\mathbf{q}_n \leftarrow \text{InitState}()$
4:     $\hat{\mathbf{x}}_{k|k} \leftarrow \text{Add}(\mathbf{q}_n)$
5:     $(\mathbf{P}_{k|k}, \mathbf{Q}, \mathbf{R}) \leftarrow \text{Initialize}(n)$
6:     **loop**
7:          $\mathbf{u}_k \leftarrow \text{GetInputSignal}(n)$
8:          $\mathbf{y}_k \leftarrow \text{GetSensorData}(n)$
9:          $\mathbf{A}_k \leftarrow \frac{\partial f(\mathbf{x}_k, \mathbf{u}_k)}{\partial \mathbf{x}_k}$
10:        $\mathbf{C}_k \leftarrow \frac{\partial g(\mathbf{x}_k)}{\partial \mathbf{x}_k}$
11:        $(\hat{\mathbf{x}}_{k+1|k+1}, \mathbf{P}_{k+1|k+1}) \leftarrow \text{EKF}(\hat{\mathbf{x}}_{k|k}, \mathbf{P}_{k|k}, \mathbf{u}_k, \mathbf{y}_k, \mathbf{A}_k, \mathbf{C}_k)$
12:        $\hat{\mathbf{y}}_{k|k} \leftarrow \text{Equation (11.9)}$
13:        $\mathbf{e} \leftarrow \text{DetermineError}(\hat{\mathbf{y}}_{k|k}, \mathbf{y}_k)$
14:        $k \leftarrow k + 1$
15:        **if** $\text{norm}(\mathbf{e}) < \delta$ **and** $n < N$ **then**
16:           **break**
17:        **end if**
18:      **end loop**
19:     $n \leftarrow n + 1$
20: **end loop**

A quadratic potential analogous with the spring-mass system is used to guide the end effector to a desired configuration $\mathbf{q}_d$ as follows

$$U_d(\mathbf{q}) = \frac{1}{2}k(\mathbf{q} - \mathbf{q}_d)^T(\mathbf{q} - \mathbf{q}_d), \tag{11.10}$$

where k stands for a positive constant. The use of the kinematic model of continuum manipulator leads to the modification of a generalized vector field $\mathbf{F}$ into the task-space velocity of the manipulator $\dot{\mathbf{p}}$ as follows

$$\dot{\mathbf{p}}_{\mathbf{p}_d} = -\nabla U_d(\mathbf{p}) = -k(\mathbf{p} - \mathbf{p}_d). \tag{11.11}$$

This task-space velocity $\dot{\mathbf{p}}$ can then be mapped into actuator-space velocity $\dot{\mathbf{q}}$ via Equation (11.3). The vector field used to ensure the manipulator body safe from collision will be explained in more detail.

11.3.1 Electric-field-based Navigation

In the first part of this section, we present the modified version of the classical electric-field-based navigation based on our previous works in [18] to do obstacle avoidance. Each obstacle will produce a repulsive vector field as follows

$$\dot{\mathbf{p}}_{\mathcal{O}} = \begin{cases} \eta(\frac{1}{\rho} - \frac{1}{\rho_0})\frac{1}{\rho^2}\frac{\partial\rho}{\partial\mathbf{p}} & \text{if } \rho < \rho_0 \\ 0 & \text{if } \rho \geq \rho_0 \end{cases}. \tag{11.12}$$

$\rho = \sqrt{(\mathbf{p} - \mathbf{p}_{\mathcal{O}})^T(\mathbf{p} - \mathbf{p}_{\mathcal{O}})}$ represents the closest distance between the point on manipulator's body and the obstacle surface, while η and ρ_0 stand for positive constant and the limit distance of influence, respectively.

Several points along the manipulator body, called "point subjected to potentials" (PSP), are chosen to be a working point of the proposed vector field to ensure that the body of the manipulator is safe from collision. The pose of a PSP in segment-i is given by $\mathbf{p}(\mathbf{q}, \xi = \lambda_i)$, where

$$\lambda_i = \begin{bmatrix} \xi_1 = 1 & \cdots & \xi_i \in [0,1] & \xi_{i+1} = 0 & \cdots & \xi_N = 0 \end{bmatrix}^T. \tag{11.13}$$

It is noted that only the closest pair of obstacle-PSP will contribute to the motion at a time. Only manipulator's end effector will be influenced by the combined attractive and repulsive vector fields while the rest of the body is influenced only by the obstacle repulsion. The task-space velocities of the end effector and the closest PSP to the obstacle are then transformed to the actuator space, manifested as an input signal for the kinematic model and the pose estimation, $\mathbf{u}_k$.

$$\mathbf{u}_k = \dot{\mathbf{q}} = \mathbf{J}_e^+ \dot{\mathbf{p}}_{p_d} + \mathbf{J}_a^+ \dot{\mathbf{p}}_\mathcal{O}. \tag{11.14}$$

Here the terms $\mathbf{J}_e$ and $\mathbf{J}_a$, respectively, represent the end-effector's Jacobian and the PSP Jacobian, $\dot{\mathbf{p}}_\mathcal{O}$ is a repulsive vector field generated by the closest obstacle to the PSP, while the ($^+$) operation is defined as $\mathbf{J}^+ = \mathbf{J}^T(\mathbf{J}\mathbf{J}^T)^{-1}$ [4].

11.3.2 Magnetic-field-based Navigation

One of the fundamental formulas on the magnetic field generation is the Biot-Savart equation describing the magnetic field generated by a current-carrying wire. An electrical current i_o flowing on the wire of infinitesimal length $\mathbf{dl}_o$ as illustrated in Figure 11.3.2 will generate an infinitesimal magnetic field $\mathbf{dB}$ given by the following equation [25]

$$\mathbf{dB} = \frac{\mu_0}{4\pi} \frac{i_o \mathbf{dl}_o \times \mathbf{r}}{|\mathbf{r}|^3}. \tag{11.15}$$

Here, $\mathbf{r}$, μ_0, and $\times$, respectively, stand for the position of a point with respect to the wire, the permeability constant, and the vector cross-product operation. This magnetic field will produce a force $\mathbf{dF}$ on any other current-carrying wire with an infinitesimal length $\mathbf{dl}_a$ and current i_a, as depicted in Figure 11.3.2, as described by the following equation

$$\mathbf{dF} = i_a \mathbf{dl}_a \times \mathbf{B}. \tag{11.16}$$

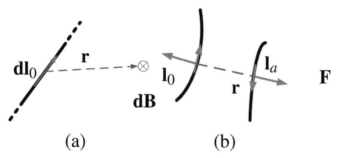

(a)   (b)

Figure 11.3 (a) An illustration of current-carrying wire with electrical current flows in the direction of $\mathbf{dl}_o$ inducing magnetic field $\mathbf{dB}$ which points inside toward the paper. (b) Two current-carrying wires with currents flow opposite to each other, $\mathbf{l}_a$ and $\mathbf{l}_o$, will produce repulsion force $\mathbf{F}$ on both wires.

From the previous equations, we can then derive the force interaction between two current-carrying wires. After some simplification and dropping all the infinitesimal notation, we can write down the interaction force as follows

$$\mathbf{F} = \frac{\mu_0 i_a i_o}{4\pi} \frac{\mathbf{l}_a \times (\mathbf{l}_o \times \mathbf{r})}{|\mathbf{r}|^3}. \tag{11.17}$$

Inspired by this physical phenomena, we can see continuum manipulator as a current-carrying wire with current direction $\mathbf{l}_a$ which will induce artificial current vector $\mathbf{l}_o$ on the obstacle surface located at position $\mathbf{r}_o$ with respect to the manipulator. This induced current on the obstacle will produce magnetic field $\mathbf{B}$ which will produce force on the robot current in such a way that the force $\mathbf{F}$ will avoid the robot body from colliding with the obstacle. Using the fact that $\mathbf{r}_o = -\mathbf{r}$ by definition, Equation (11.17) can be expressed as follows

$$\mathbf{F}(\mathbf{r}_o) = c \ \mathbf{l}_a \times (\mathbf{r}_o \times \mathbf{l}_o) \ f(|\mathbf{r}_o|). \tag{11.18}$$

Here, $c > 0$ and $f(|\mathbf{r}_o|) \geq 0$ denote a scalar constant and positive scalar function, respectively. To simplify the cross-product operation, we introduce a skew-symmetric matrix $\hat{\mathbf{l}}$ as a substitution to the vector cross-product operation $\mathbf{l} \times$ of vector $\mathbf{l} = \begin{bmatrix} l_x & l_y & l_z \end{bmatrix}^T$ defined as

$$\hat{\mathbf{l}} = \begin{bmatrix} 0 & -l_z & l_y \\ l_z & 0 & -l_x \\ -l_y & l_x & 0 \end{bmatrix}. \tag{11.19}$$

The magnetic-field-based navigation extends the formulation presented in Equation (11.18). An artificial current i_a is assumed to flow continuously along the body of the continuum manipulators from the base to the tip. At every point on the robot's body, the current direction, specified by $\mathbf{l}_a$, is always tangential to the robot's curvature as illustrated in Figure 11.4. Once the parts of the robot body are close enough to the obstacle, the closest obstacle point to the robot will induce a shadow current i_o with the current direction $\mathbf{l}_o$. We designed this current direction $\mathbf{l}_o$ in such a way that the interaction force results in a repulsive behavior. For that purpose, we take inspiration from the repulsion phenomena observed in the case of two parallel wires carrying current in the opposite direction as shown in Figure 11.3.2. Hence, the current direction on the obstacle is defined as

$$\mathbf{l}_o = -\mathbf{l}_a. \tag{11.20}$$

The characteristic of the produced force will be described further in the following lemmas.

Figure 11.4 The definition of the current direction $\mathbf{l}_a$, drew as orange arrows, on the manipulator's body.

Lemma 1. *For every obstacle point with current direction $\mathbf{l}_o$ as defined in Equation (11.20) and located at position $\mathbf{r}_o$ with respect to the robot, vector $\mathbf{l}_o$, $\mathbf{l}_a$, and $\mathbf{r}_o$ are coplanar.*

Proof. Three vectors are coplanar if and only if they are not linearly independent, i.e., we can write down one vector as a linear combination of two other vectors. We can write down

$$\mathbf{l}_o = a \ \mathbf{l}_a + b \ \mathbf{r}_o, \qquad (11.21)$$

where a and b are scalar. If we choose $a = -1$ and $b = 0$, the above equation will simply become Equation (11.20), thus proofing the lemma. □

Lemma 2. *For every obstacle point with current direction $\mathbf{l}_o$ as defined in Equation (11.20) and located at position $\mathbf{r}_o$ with respect to the robot, the produced force defined in Equation (11.18) will always be repulsive except when $\mathbf{l}_o$, $\mathbf{l}_a$, and $\mathbf{r}_o$ lie on the same line.*

Proof. Lemma 1 concludes that $\mathbf{l}_o$, $\mathbf{l}_a$, and $\mathbf{r}_o$ are coplanar; hence, each vector can be described using a pair of unit vectors of this plane: $\mathbf{l}_a = \begin{bmatrix} l_{ax} & l_{ay} \end{bmatrix}^T$, $\mathbf{r}_o = \begin{bmatrix} r_{ox} & r_{oy} \end{bmatrix}^T$. Assuming $cf(|\mathbf{r}_o|) = 1$, the produced force on the robot $\mathbf{F}$ can be derived from Equations (11.18) to (11.20) as

$$\mathbf{F} = \begin{bmatrix} -l_{ay}^2 r_{ox} + l_{ax} l_{ay} r_{oy} \\ -l_{ax}^2 r_{oy} + l_{ax} l_{ay} r_{ox} \end{bmatrix}. \qquad (11.22)$$

A dot-product between the force $\mathbf{F}$ and position vector of obstacle $\mathbf{r}_o$ can be simplified as follows

$$\mathbf{F}^T \mathbf{r}_o = 2 l_{ax} l_{ay} r_{ox} r_{oy} - l_{ay}^2 r_{ox}^2 - l_{ax}^2 r_{oy}^2 = -(l_{ay} r_{ox} - l_{ax} r_{oy})^2. \qquad (11.23)$$

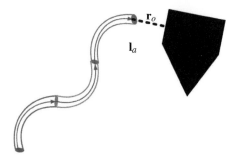

Figure 11.5　Configuration where the standard continuous current algorithm fails, i.e., when the current l_a and vector $\mathbf{r}_o$ toward polygonal obstacle illustrated as a black object lie on the same line.

From Equation (11.23), it can be concluded that $\mathbf{F}^T\mathbf{r}_o \leq 0$. This implies that in the case of $\mathbf{F}^T\mathbf{r}_o < 0$, the force $\mathbf{F}$ will always have component in the opposite direction of $\mathbf{r}_o$, thus, producing a repulsive behavior (away from the obstacle). The exception occurs when $\mathbf{F}^T\mathbf{r}_o = -l_{ay}r_{ox} + l_{ax}r_{oy} = 0$, which is the implication of $\hat{\mathbf{l}}_a\mathbf{r}_o = \mathbf{0}$. It means that $\mathbf{r}_o$ is in the same or opposite direction of l_a, which will occur only when l_a (and consequently l_o) lies on the same line as $\mathbf{r}_o$.　□

Lemma 3. *The produced force $\mathbf{F}$ in Equation (11.18) has no component along the line tangential to the continuum manipulator's curvature.*

Proof. Suppose we define $\mathbf{B}$ as follows

$$\mathbf{B} = c(\mathbf{r}_o \times \mathbf{l}_o)\ f(|\mathbf{r}_o|),\tag{11.24}$$

the dot product between l_a and $\mathbf{F}$ can be written as

$$\mathbf{l}_a^T\mathbf{F} = \mathbf{l}_a^T\hat{\mathbf{l}}_a\mathbf{B}.\tag{11.25}$$

From the definition of the skew-symmetric matrix in Equation (11.19), it can be shown that $\mathbf{l}_a^T\hat{\mathbf{l}} = \mathbf{0}^T$ and thus it produces a zero product between l_a and $\mathbf{F}$, meaning that the two vectors are perpendicular. By choosing l_a to flow along the curvature of the continuum manipulator, it can be concluded that the force $\mathbf{F}$ will have no component in this direction.　□

This proposed algorithm is advantageous for the continuum manipulator due to the fact that it has no force component in the direction tangential to the manipulators curvature, as has been proven in Lemma 3. Hence, it is suitable for manipulators with more dominant bending than extension ability, like the case for most of the current designs [10]. Compared to the works proposed in [18], where several points along the backbone of the manipulator are used as subject to the repulsive field, this method is superior since it will have little contribution to the manipulator's extension, and thus, help avoiding length constraint in such a manipulator.

However, this method as it also has a drawback. As mentioned in Lemma 2, it has zero force for the case when the robot current $\mathbf{l}_a$ and the obstacle relative position vector $\mathbf{r}_o$ lie on the same line. This will happen when an obstacle is located at exactly above the distal tip, as illustrated in Figure 11.5. Although this practically never happens in reality, in order to make sure that it is solved, we can add a special condition when $\hat{\mathbf{l}}_a\mathbf{r}_o = \mathbf{0}$. In this case, the current direction $\mathbf{l}_o$ can be chosen as any $90°$ rotation of robot current direction $\mathbf{l}_a$.

Finally, the scalar function $f(|\mathbf{r}_o|)$ can be chosen to have the following form

$$f(|\mathbf{r}_o|) = \begin{cases} \left(\dfrac{1}{|\mathbf{r}_o|} - \dfrac{1}{r_b}\right)\dfrac{1}{|\mathbf{r}_o|^2} & \text{if } |\mathbf{r}_o| < r_b \\ 0 & \text{if } |\mathbf{r}_o| \geq r_b \end{cases}, \qquad (11.26)$$

where r_b specifies a limit distance of the potential influence. Hence, the proposed method has been fully described.

11.3.3 The Complete Algorithm

The pose estimation stage is combined to the navigation stage as illustrated in Figure 11.6. Here, the pose estimation stage provides information on the end-effector pose estimate $\mathbf{p}(\mathbf{x}_k, \xi = \chi_N)$ along with the pose estimate of closest point on the robot body to the obstacle $\mathbf{p}(\mathbf{x}_k, \xi = \lambda_i)$ at every iteration for the navigation purpose. This is done by applying the forward kinematics equation to the state estimate from the EKF as follows

$$\hat{\mathbf{p}}_k(\xi) = \mathbf{p}(\hat{\mathbf{x}}_{k|k}, \xi). \qquad (11.27)$$

The produced control signal $\mathbf{u}_k$ from the navigation stage is then used as an input to the estimation stage to estimate the state value for the next iteration. Lastly, as explained in more detail in our previous work [23], due to the early error of the estimation stage, the navigation stage will be activated only when a certain condition relating to the tip position estimation errors for all segments is fulfilled. In short, we have combined the pose estimation and

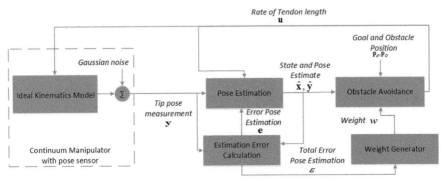

Figure 11.6   The way the pose estimator is combined with the obstacle avoidance algorithm.

navigation algorithm to ensure collision-free movement of the manipulator's body.

11.4 Results and Discussion

Some simulation results to confirm the performance of the algorithm are presented in this section. All presented simulations use the Robot Operating System (ROS) framework. The model of a three-segment continuum manipulator was used to test the algorithm. Five points per segment, distributed uniformly along the backbone of each segment, are chosen to be the working point of the navigation field. A spherical obstacle with varying dimension and behavior is placed in the vicinity of the manipulator to test the proposed method. For the pose estimation stage, the measurement data of the tip pose y_k of each segment are generated via ideal kinematic model added with Gaussian noise to replace a 3-DOF position tracker assumed to be embedded in the tip of each segment. A more detail description on the value of parameters used in the pose estimation and electric-field-based navigation can be referred back to our work in [23] and [18].

The first part of simulation shows the performance of the pose estimation stage in the absence of obstacle. In this case, a circular trajectory is used as a reference path for the end effector. We can see from Figures 11.7a and 11.7b how the multistage EKF is able to cope with the varying input signal and sensory data and successfully track the real state value.

In the second part of simulation, a moving spherical obstacle with varying size is assumed to move in the proximity of the manipulator body. For all

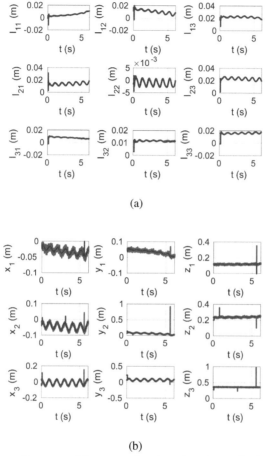

(a)

(b)

Figure 11.7 The performance of the pose estimation when the manipulator's end effector is given a circular trajectory shows (a) the true state (red line) and the estimated state (blue line) comparison and (b) the true pose with Gaussian noise (red line) and the estimated pose (blue line) comparison, respectively.

figures, the order of the subfigures is from the upper left picture to the upper right picture and then the lower left picture to the lower right picture.

We first implemented the electric-field-based navigation. Here, we assume two cases: the one in which the obstacle is close to the bottom segment drawn in red as shown in Figure 11.8 and the one in which the obstacle is close to the middle segment drawn in green as shown in Figure 11.9.

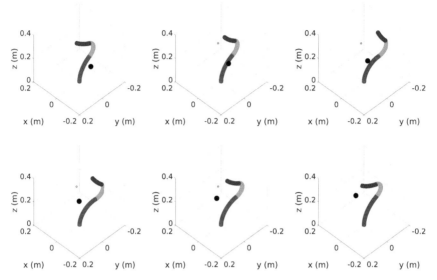

Figure 11.8   The movement of a three-segment continuum manipulator avoiding moving obstacle drawn in black in the proximity of the bottom segment drawn in red. The order of subfigures is from the upper left picture to the upper right picture and then the lower left picture to the lower right picture.

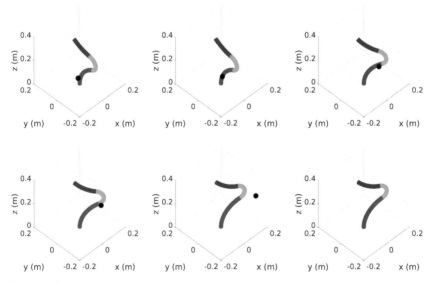

Figure 11.9   The movement of a three-segment continuum manipulator avoiding moving obstacle drawn in black in the proximity of the middle segment drawn in green. The order of subfigures is from the upper left picture to the upper right picture, and then the lower left picture to the lower right picture.

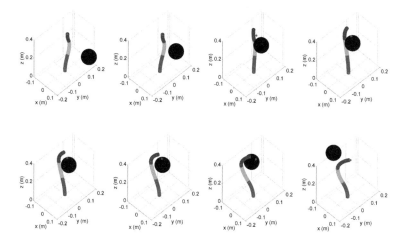

Figure 11.10 Simulation of the magnetic-field-based navigation working to avoid moving obstacle drew in black sphere. The order of movement is from the upper left to the upper right and then the lower left to the lower right.

The results demonstrate how the combined pose estimation and electric-field-based navigation is able to make the body of the manipulator safe from colliding with unknown and even dynamic spherical obstacle.

We then implement the magnetic-field-based navigation to the robot model. In the first scenario, shown in Figure 11.10, the desired end-effector position is fixed while the obstacle, with 5 cm radius, is able to move in the vicinity of the manipulator's body. While, in the second scenario, shown in Figure 11.11, the obstacle, with 3 cm radius, is made to be fixed while the end effector starts from a position located at some distance from the target point. After the end effector reaches the target, the target's position is moved immediately so that the manipulator starts to move to the new target position.

In Figure 11.10, we can see that the spherical obstacle drew in black moves toward the manipulator's body. The force produced by the magnetic interaction of current flowing along the backbone of the manipulator and shadow current flowing on the surface of the obstacle produces a repulsive behavior which makes the manipulator body moves away from the approaching obstacle. We can see that the closest point on the manipulator to the obstacle tends to move in the direction opposite of the vector connecting the robot and the obstacle, as has been proven in Lemma 2. The figure also shows how the manipulator behaves in such a way that the bending

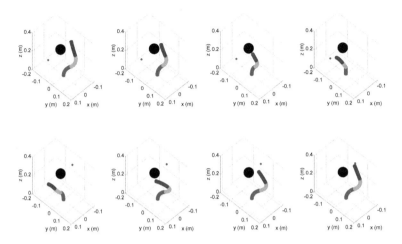

Figure 11.11   Simulation of the magnetic-field-based navigation working to avoid static obstacle (black sphere) while moving toward the goal (red dot). The order of movement is from the upper left to the upper right and then the lower left to the lower right.

movement is more dominant than the extension, as can be observed that the segments' lengths do not change a lot during the whole movement. This has also been proven in Lemma 3 and becomes a major advantage of this method to be implemented for the continuum manipulator compared to the standard potential field. Figure 11.11 shows how the manipulator can avoid the static obstacle on the way toward the target (shown in the upper figures). This is also the case when the target is moved back to the end-effector's initial position (shown in the lower figures).

11.4.1 Discussion

From the simulation results, it can be seen that in general both the electric-field-based navigation and the magnetic-field-based navigation work success-fully in avoiding the obstacle. The implementation of the multistage EKF as a pose estimator is useful to estimate the pose of any point along the manipulator's body such that collision with surrounding the obstacle can be avoided. The reactive nature of the navigation algorithm also enables the manipulator to avoid collision with dynamically moving and unknown obstacle without prior information of the surrounding environment. The magnetic-field-based navigation is also superior due to the fact that it is designed to comply with the continuum manipulator's limitation especially on its extension capability.

However, the repulsive nature of the algorithm makes it still susceptible to local minima when the manipulator moves in a complex environment. This problem is left to be investigated further in future works.

11.5 Conclusion

In this article, we present our recent works on reactive navigation for the case of the continuum manipulator. The navigation problem we are trying to deal with in this chapter is that in which the information of the environment is unknown prior to the robot movement, which makes the robot have to rely on local sensory information and response in a reactive manner. The two presented algorithms, one inspired by the electric field phenomena while the other inspired by the magnetic field, are shown to successfully avoid the continuum manipulator body from colliding with the environment. The magnetic-field-based navigation is shown to be more suitable for navigating the class of the continuum manipulator robot due to the fact that it mainly works to bend, rather than extend, the segment of manipulator. In the future, the practical implementation of the algorithm will be explored along with the effort to overcome the current limitations. Implementation of the method on other continuum manipulators such as a snake-like robot, or a more complex dynamic model of the continuum manipulator will also be investigated.

Acknowledgment

Research is partially supported by King's College London, the STIFF-FLOP project grant from the European Commission Seventh Framework Programme under grant agreement 287728, the Four By Three grant from the European Framework Programme for Research and Innovation Horizon 2020 under grant agreement no 637095, and the Indonesia Endowment Fund for Education, Ministry of Finance Republic of Indonesia.

References

[1] McMahan, W., Chitrakaran, V., Csencsits, M., Dawson, D., Walker, I. D., Jones, B. A., et al. (2006). "Field trials and testing of the OctArm continuum manipulator," in *Proceedings of the IEEE International Conference on Robotics and Automation*, 2336–2341.

[2] Kim, Y.-J., Cheng, S., Kim, S., and Iagnemma, K. (2013). "A novel layer jamming mechanism with tunable stiffness capability for minimally invasive surgery," in *Proceedings of the IEEE Transactions on Robotics*, 29, 1031–1042.

[3] Lyons, L. A., Webster, R. J., and Alterovitz, R. (2009). "Motion planning for active cannulas," in *Proceedings of the IEEE/RSJ International Conference on Intelligent Robots and Systems*, 801–806.

[4] Maghooa, F., Stilli, A., Noh, Y., Althoefer, K., and Wurdemann, H. A. (2015). "Tendon and pressure actuation for a bio-inspired manipulator based on an antagonistic principle," in *Proceedings of the 2015 IEEE International Conference on Robotics and Automation (ICRA)*, 2556–2561.

[5] Qi, P., Qiu, C., Liu, H., Dai, J. S., Seneviratne, L., and Althoefer, K. (2014). "A novel continuum-style robot with multilayer compliant modules," in *Proceedings of the IEEE/RSJ International Conference on Intelligent Robots and Systems*, 3175–3180.

[6] Mahl, T., Hildebrandt, A., and Sawodny, O. (2014). "A variable curvature continuum kinematics for kinematic control of the bionic handling assistant," in *Proceedings of the IEEE Transactions on Robotics*, 30, 935–949.

[7] Cianchetti, M., Ranzani, T., Gerboni, G., De Falco, I., Laschi, C., and A. Menciassi. (2013). "STIFF-FLOP surgical manipulator: mechanical design and experimental characterization of the single module," in *Proceedings of the IEEE/RSJ International Conference on Intelligent Robots and Systems*, 3576–3581.

[8] Niu, G., Zheng, Z., and Gao, Q. (2014). "Collision free path planning based on region clipping for aircraft fuel tank inspection robot," in *Proceedings of the IEEE International Conference on Robotics and Automation*, 3227–3233.

[9] Torres, L. G. and Alterovitz, R. (2011). "Motion planning for concentric tube robots using mechanics-based models," in *Proceedings of the IEEE/RSJ International Conference on Intelligent Robots and Systems*, 5153–5159.

[10] Walker, I. D., Carreras, C., McDonnell, R., and Grimes, G. (2006). Extension versus bending for continuum robots. *Int. J. Adv. Robot. Syst.* 3, 171–178.

[11] Webster, R. J., III, and Jones, B. A. (2010). Design and kinematic modeling of constant curvature continuum robots: a review. *Int. J. Rob. Res.* 29, 1661–1683.

[12] Xiao, J., and Vatcha, R. (2010). "Real-time adaptive motion planning for a continuum manipulator," in *Proceedings of the IEEE/RSJ International Conference on Intelligent Robots and Systems*, 5919–5926.

[13] Xu, J., Duindam, V., Alterovitz, R., and Goldberg, K. (2008). "Motion planning for steerable needles in 3D environments with obstacles using rapidly-exploring random trees and backchaining," in *Proceedings of the IEEE international conference automation science engineering*, 41–46.

[14] Duindam, V., Alterovitz, R., Sastry, S., and Goldberg, K. (2008). "Screw-based motion planning for bevel-tip flexible needles in 3D environments with obstacles," in *Proceedings of the IEEE International Conference on Robotics and Automation*, 2483–2488.

[15] Chen, G., Pham, M. T., and Redarce, T. (2009). Sensor-based guidance control of a continuum robot for a semi-autonomous colonoscopy. *Robot. Auton. Syst.* 57, 712–722.

[16] Godage, I. S., Branson, D. T., Guglielmino, E., and Caldwell, D. G. (2012). "Path planning for multisection continuum arms," in *Proceedings of the International Conference on Mechatronics and Automation (ICMA)*, 1208–1213.

[17] Khatib, O. (1985). "Real-time obstacle avoidance for manipulators and mobile robots," in *Proceedings of the IEEE International Conference on Robotics and Automation*, 2, 500–505.

[18] Ataka, A., Qi, P., Liu, H., and Althoefer, K. (2016). "Real-time planner for multi-segment continuum manipulator in dynamic environments," in *Proceedings of the IEEE International Conference Robotics and Automation*, 4080–4085.

[19] Ataka, A., Qi, P., Shiva, A., Shafti, A., Wurdemann, H., Dasgupta, P., et al. (2016). "Towards safer obstacle avoidance for continuum-style manipulator in dynamic environments," in *Proceedings of the 2016 6th IEEE International Conference on Biomedical Robotics and Biomechatronics (BioRob)*, 600–605.

[20] Singh, L., Stephanou, H., and Wen, J. (1996). "Real-time robot motion control with circulatory fields," in *Proceedings of the IEEE International Conference on Robotics and Automation*, 3, 2737–2742.

[21] Singh, L., Wen, J., and Stephanou, H. (1997). "Motion planning and dynamic control of a linked manipulator using modified magnetic fields," in *Proceedings of the 2004 IEEE International Conference on Control Applications*, 9–15.

[22] Haddadin, S., Belder, R., and Albu-Schäffer, A. (2011). "Dynamic motion planning for robots in partially unknown environments," in *Proceedings of the IFAC World Congress*, 18.

[23] Ataka, A., Qi, P., Shiva, A., Shafti, A., Wurdemann, H., Dasgupta, P., et al. (2016). "Real-time Pose estimation and obstacle avoidance for multi-segment continuum manipulator in dynamic environments," in *Proceedings of the 2016 IEEE/RSJ International Conference on Intelligent Robots and Systems (IROS)*, 2827–2832.

[24] Siciliano, B. (1990). Kinematic control of redundant robot manipulators: a tutorial. *J. Intell. Robot. Syst.* 3, 201–212.

[25] Jackson, J. D. (1999). *Classical electrodynamics*, 3rd edition. (New York, NY: Wiley).

[26] Bergeles, C., and Dupont, P. E. (2013). "Planning stable paths for concentric tube robots," in *Proceedings of the IEEE/RSJ International Conference on Intelligent Robots and Systems*, 3077–3082.

[27] Palmer D., Cobos-Guzman, S., and Axinte, D. (2014). Real-time method for tip following navigation of continuum snake arm robots. *Robot. Auton. Syst.* 62, 1478–1485.

PART IV

Human Interface

12

The Design of a Functional STIFF-FLOP Robot Operator's Console

Łukasz Mucha, Krzysztof Lis, Dariusz Krawczyk and Zbigniew Nawrat

Prof. Z. Religa Foundation of Cardiac Surgery Development, Zabrze, Poland

Abstract

The chapter presents the development of a human–machine interface developed within the framework of the STIFF-FLOP project. The workstation concept of a surgeon is based on the assumption that the tool control method is consistent with the natural work of the surgeon. An integrated central unit (console) prototype was developed with several channels of bi-directional information flow between the robot and an operator. The STIFF-FLOP control console – Robin Heart – is equipped with two monitors, a haptic device with force feedback, and a foot-controlled button.

12.1 Introduction

The STIFF-FLOP robot consists of mechanical units, actuators, sensors, effectors, and a control unit. An important element of the proposed surgeon–machine interface is the introduction of the haptic feedback which provides the surgeon with sensations of touch and resistance to the tool tip, as they operate inside the body (Figure 12.1).

The Foundation of Cardiac Surgery Development (FCSD) team developed and implemented a control system of the console according to the requirements of medical systems and their safety (robust systems), as well as optimized the system of communication between the operator (surgeon) and the robot [1–3].

The specificity of the task lies in the limited space for the tool manipulation inside the human body. The user interface must not only meet the criteria of technical efficiency, but also accommodate the skills and experience of the

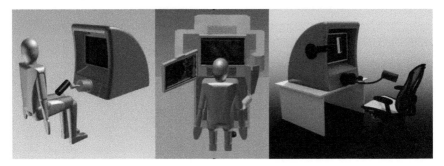

Figure 12.1   STIFF-FLOP console – design concept.

surgeon, typically acquired during the use of classical or laparoscopic tools. The moves made by the surgeon using the motion controller are mirrored by the robot. While during operations with a surgical tool (such as opening and closing scissors), as well as adding extra degrees of freedom and additional tasks (e.g., coagulation) of the end effector is performed by pressing a specific button or changing the position of the input controller.

The console model developed consists of a user interface, which allows the operator to control the position of the end effector. This is implemented by monitoring position and force feedback, obtained during the interaction between the robot and the environment. The action of the robot is visualized on a set of monitors in an ergonomic position for the operator. The operator can monitor and select the individual control parameters, such as scaling the motion, force, or the number of active sensors (Figure 12.1) [4–6].

12.2 Design of Improved Haptic Console

The control console is based on a user interface. The user controls the position of the end effector of the STIFF-FLOP robot arm. The haptic feedback information regarding the position and the operating forces is obtained during robot tool–environment interaction, using sleeves integrated with pneumatic or vibrating actuators to relay collisions between the robot's arm and the environment to the surgeon. The main console concept and the technical structure of the operator control unit are presented in Figure 12.2. Additionally, the operator can use a touchpad to set certain control parameters, such as scaling the motion or force, as well as adjusting the number of active sensors. The user interface RiH Delta *RobinHand F* haptic console was designed and developed using FCSD's experience from the previously conducted Robin Heart projects (Chapter 16): the interface allows the digital mapping of the

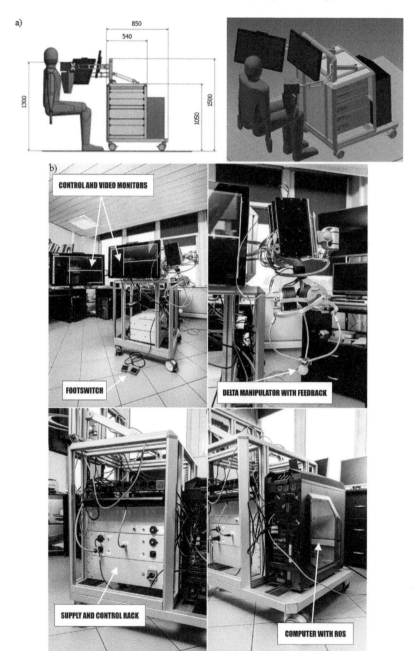

Figure 12.2 STIFF-FLOP console – (a) sketches of design concept and (b) modular prototype with components' description.

free movement of the surgeon's hands and lets the surgeon feel the force impact from the working tip (tools) of the robot operating in the environment of the patient by transferring it to his or her palm [7].

The STIFF-FLOP console includes:

1. Computer unit – with Linux and ROS (Robot Operating System);
2. Mechanical structure of the console body made from aluminum profiles' system;
3. Power supply and control rack system;
4. Two monitors;
5. Delta haptic device;
6. Haptic sleeve integrated with pneumatic/vibrating actuators (Chapter 14).

The STIFF-FLOP control console has a modular design based on a modular (rack) system. There are three different modules: power rack, vibro rack, and haptic rack (Figure 12.3).

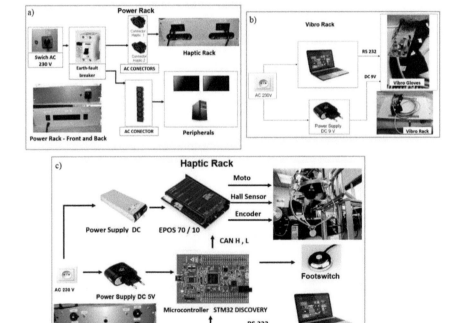

Figure 12.3 STIFF-FLOP console – Rack system: (a) power module diagram, (b) vibro module diagram, and (c) haptic module diagram.

12.2.1 Second Version of STIFF-FLOP Console

Further on, a second version of the console was developed. In order to reduce the dimensions of the console (Figure 12.4), the FCSD console has been modernized. In the first version of the console, desktop PCs are used to interface with the system using ROS. The amount of data processed was evaluated. To introduce a more cost-effective solution, a modular console using Raspberry Pi microcomputers was developed. The Robin STIFF-FLOP console has two of Raspberry Pi's microcomputers with ROS on board. Communication is established via TCP/IP protocol. One of the microcomputers communicates with the microcontrollers of the motion controller, while the second one controls the actuators of the robot. The whole system is designed to control the motion of the actuators, to analyze data from the sensors, and to control the force feedback.

The haptic control unit precisely determines the location and orientation of the tool tip in the three-dimensional space. The haptic control unit collects information about the positions of the arms of the integrated Delta interface, which is determined based on the voltage on the robot's motors. Motor controllers collect this information and convert it into position values in the Cartesian space. The reference point (origin) of the controller (0, 0, 0) was assumed for the minimum position on the Z-axis.

The microcontroller performs data acquisition of the force vector resultant from the interaction between the tissue and the STIFF-FLOP manipulator

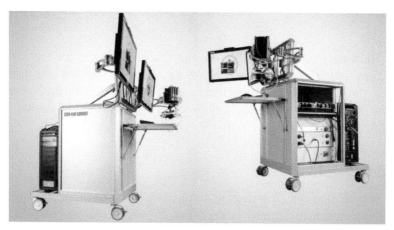

Figure 12.4   STIFF-FLOP console made by the Foundation of Cardiac Surgery Development (FCSD) (first version).

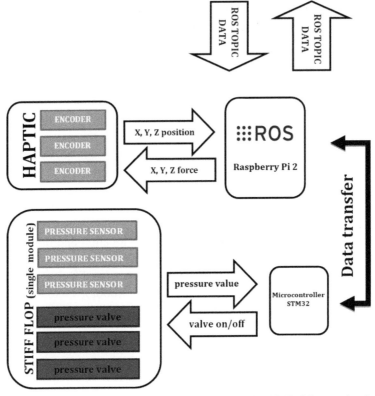

Figure 12.5   Control system (console and haptic) for the STIFF-FLOP arm, developed by the FCSD.

from the ROS environment. Further on, the force vector is transmitted to the haptic console. The console is ready to accommodate a vibration glove to enhance the haptic feedback on the surgeon's hand. The control structure of the STIFF-FLOP console for the STIFF-FLOP arm is shown in Figure 12.5.

12.3 Conclusion

The motion controller is a separate electromechanical system converting intuitive hand movements to continuous digital signal which enables controlling the robot's tools. The system may be equipped with modules which record, process, and transmit feedback data to the operator. The haptic feedback reflects, in a number of ways, the manner of interaction (force, optical, thermal, vibratory, etc.) between the tool/the entire arm and objects within the operating field. Signals carrying information about the operator's movements as well as feedback signals may be scaled and filtered, which is a great advantage of tele-manipulated devices. The control system of such devices must ensure the required accuracy and resolution of movement, rescaling the range of hand movements to the range of the robot's arms movement, and eliminate the effect of the operator's shaking hands.

It is assumed during console development that the tool tip actuation is performed intuitively and that the robot operator receives both visual and haptic feedback, which enables accurate control of the robot. The challenge of the STIFF-FLOP project was to allow the control of the tool tip mounted on an arm with a variable, controllable stiffness, and geometry. The operator of this type of robot, in addition to supervising the tool, should also have information about the position of the flexible, octopus-inspired arm. A set of sensors on the surface of the tool's arm (described in Part II of this book) transmits the information regarding contact or collision with other elements. For this reason, the FCSD team has developed a specially equipped console – an ergonomic surgeon's workstation. The console has been technically tested in the laboratory environment, as well as during experimental tele-operations – when the robot was remotely controlled from various distances. The work on the optimization of the console and the motion controllers, as well as the development of suitable software depending on the type of robot, tool, and surgical procedure are continued by the Zabrze team [6–11].

Acknowledgments

The project of flexible tool was supported in part by the European Commission within the STIFF-FLOP FP7 European project FP7/ICT-2011-7-287728. The authors thank K. Rohr and M. Jakubowski from the Prof. Z. Religa Foundation of Cardiac Surgery Development for their assistance with the technical aspects of the study.

References

[1] Nawrat, Z., Kostka, P., Dybka, W., Rohr, K., Podsędkowski L., Śliwka J., et al. (2010). Pierwsze eksperymenty na zwierzętach robota chirurgicznego Robin Heart. *PAR* 2, 539–545.

[2] Nawrat, Z., Kostka, P., and Małota, Z. (2011). Ergonomiczne stanowisko operatora robota chirurgicznego Robin Heart – prace projektowe, konstrukcyjne I badawcze 2009–2010. *PAR* 2, 619–626.

[3] Nawrat, Z., Dybka, W., Kostka, P., and Rohr, K. (2008). Konsola sterowana robotem chirurgicznym Robin Heart. *PAR* 708–718.

[4] Greer, A. D., Newhook, P. M., and Sutherland, G. R. (2008). Human–machine interface for robotic surgery and stereotaxy," in *Proceedings of the IEEE/ASME Transactions on Mechatronics*, 13.

[5] Heng, P., Cheng, C., Wong, T., Xu, Y., Chui, Y., Chan, K., and Tso, S. (2004). "A virtual-reality training system for knee arthroscopic surgery," in *Proceedings of the IEEE Transactions on Information Technology in Biomedicine*, 8.

[6] Krawczyk, D., and Kroczek, P. (2016). Integracja konsoli Robin STIFF-FLOP z robotem chirurgicznym Robin Heart Tele. *Med. Robot. Rep.* 5, 43–49.

[7] Rohr, K., Fürjes, P., Mucha, Ł., Radó, J., Lis, K., Dücsö, C., et al. (2016). Robin Heart force feedback/control system based on INCITE sensors – preliminary study. *Med. Robot. Rep.* 4:2299-7407.

[8] Lis, K., Lehrich, K., Mucha, Ł., and Nawrat, Z. (2017). Concept of application of the light-weight robot Robin Heart ("Pelikan") in veterinary medicine: a feasibility study. *Med. Weteryn.* 73, 88–91.

[9] Mucha, Ł., Nawrat, Z., Lis, K., Lehrich, K., Rohr, K., Fürjesb, P., and Dücsö, C. (2016). Force feedback control system dedicated for Robin Heart Pelikan. *Acta Bio-Optica et Inform. Inżyn. Biomed.* 22, 146–153.

[10] Mucha, Ł., Lehrich, K., Lis, K., Lehrich, K., Kostka, P., Sadowski, W., et al. (2015). Postępy budowy specjalnych interfejsów operatora robota chirurgicznego Robin Heart. *Med. Robot. Rep.* 4, 49–55.

[11] Fürjes, P., Mucha, Ł., Radó, J., Lis, K., Dücsö, C., et al. (2016). "Force feedback control system dedicated for Robin Heart surgical robot," in *Proceedings of the 30th Eurosensors, Procedia Engineering*, 168, 185–188.

13

Haptic Feedback Modalities for Minimally Invasive Surgery

Min Li[1], Jelizaveta Konstantinova[2] and Kaspar Althoefer[2]

[1]School of Mechanical Engineering, Xi'an Jiaotong University,
Xi'an, PR China
[2]School of Mechanical Engineering and Materials Science, Queen Mary
University of London ARQ (Advanced Robotics at Queen Mary), London,
United Kingdom

Abstract

Sense of touch is critical for surgeons to perform tissue palpation, and there are several tactile sensing that can be used to translate this information (Chapter 5). To overcome the loss of touch, which occurs during robotic-assisted surgical procedures, methods capable of providing partial haptic feedback and mimicking the physical interaction that takes place between surgical tools and human tissue during surgery have been proposed. This chapter introduces haptic feedback modalities for robot-assisted minimally invasive surgical platforms, such as STIFF-FLOP.

13.1 Introduction

During open surgery, which is performed using a single large incision, haptic information can be easily acquired by the surgeons using their hands. In recent years, robot-assisted minimally invasive surgery (RMIS) was widely applied. However, the physical contact between the surgeon and the soft tissue is cut off, which makes acquisition of haptic information during surgeries difficult [1]. Haptic information can be obtained if direct force feedback is provided via a surgical tele-manipulator [2]. However, most current robotic surgical systems, including da Vinci and Titan Medical Amadeus, do not provide haptic feedback.

229

Alternatively, a graphical display of soft tissue stiffness distribution can be used to show the locations of tumors [3–5]. The benefit of such an approach in comparison to direct force feedback is that it can be applied in combination with complex surgical manipulation when it is undesirable to disturb the performed manipulation. Visualization methods of stiffness distribution the TSS (tactile sensing system) [6–8] and the TIS (tactile imaging system) [9], were developed. However, the visual representation shows only the pressure distribution information within the sensing area. There is no pressure distribution map for the entire organ surface.

Pseudo-haptics is a feedback method attempting to create the illusion of actual haptic feedback through appropriately adapted visual cues without the need for the use of haptic devices [3]. The modification of speed or size of the computer cursor for simulating bumps and holes using a computer mouse and a desktop computer was proposed by previous researchers [4]. A resistance to motion when the cursor approaches an embedded hard nodule in soft tissue during sliding palpation can be simulated by reducing the ratio between the cursor displacement and the computer mouse displacement [5].

Distributed haptics has been introduced for tumor identification in MIS, for instance, the application of tactile sensors and tactile actuators [10]. However, the current application is limited by the complexity and high cost of tactile devices. Multi-fingered haptic feedback conveys more haptic information than single-point force feedback and requires less actuator elements compared to tactile haptic methods.

In this chapter, haptic feedback modalities for minimally invasive surgery are introduced, including force feedback, visual stiffness feedback, pseudo-haptic feedback, and haptic feedback actuators for fingertips.

13.2 Force Feedback

The most common haptic feedback modality is force feedback. Normally, force feedback is provided via a haptic device and conveyed via kinesthetic sensation. In this section we describe the use of force feedback for RMIS. An experimental tele-manipulator and force feedback platform were developed and validated, as part of the STIFF-FLOP project.

13.2.1 Experimental Setup to Validate the Experimental Tele-manipulator and the Force Feedback Platform

The force feedback system consisted of tele-manipulation setup (a slave robot and a master robot) and a vision system. Figure 13.1 displays the schematic

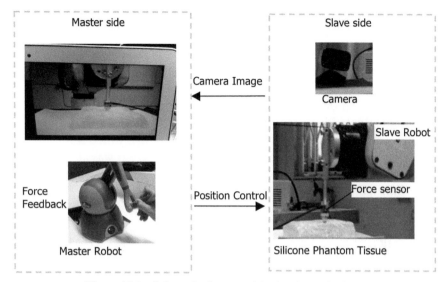

Figure 13.1 Schematic diagram of the haptic manipulator.

diagram of the system design. The right column shows the hardware at the slave side: a camera, a robot arm, a rolling indentation probe [11–13], and a silicone phantom tissue. The left column shows the configuration at the master side: live camera image and force feedback via a haptic device. The middle column lists the software used here. According to the different functions, software was classified into camera image viewer and tele-manipulation. Sensor measurements including force and position were transmitted from the slave side to enable the force feedback on the master side. Real-time images of operational site were also provided using a camera.

A master–slave tele-manipulation configuration was created to simulate the tele-manipulation environment of RMIS. PHANToM Omni (Geomagic Touch) and FANUC robot arm (M-6iB, FANUC Corporation) were used as the master robot and slave robot, respectively, with TCP/IP communication. The main loops of the master and slave sides were synchronized at a frequency of 21.3 Hz. The position of the master robot end-effector was transmitted to the salve side as a control input of the slave robot. The haptic device servo thread ran at a frequency of 1000 Hz. A force sensor was placed at the slave side. FANUC was equipped with an R-J3iC controller with embedded kinematic and dynamic controller. The sequence of the positions provided by the master (PHANToM Omni) was passed directly to the trajectory generator of the robot. In order to avoid discontinuity between the points, the trajectory generator was set to work in a linear interpolation mode following the hermite curve.

Force measurement used an ATI Nano 17 force sensor (SI-12-0.12, resolution 0.003 N with 16-bit data acquisition card). In order to rapidly scan over a large tissue area [11], a sliding indentation probe [14] was used. In order to reduce the friction during sliding over the tissue, the tissue surface was lubricated with Boots Pharmaceuticals Intrasound Gel. A force distribution matrix can be obtained using the sliding indentation probe, which shows the tissue's Young's modulus at a given indentation depth [15]. It was also found that the indentation speed did not have significant impact on the estimated elastic modulus [15].

Force feedback was applied via a haptic device according to the force sensor measurements at the slave side. The maximum executable force at nominal position of this 3-DOF of force feedback device, PHANToM Omni, was 3.3 N. Force data have three components f_x, f_y, and f_z. The perpendicular reaction force was generated from f_z. The horizontal force was generated from the resultant of f_x and f_y, and the force direction was acquired based on the difference between the previous updated position and the current position (Equation 13.1). The horizontal component vector of the force direction was then transformed to a unit vector with the same direction. The tangent force was provided in the same direction of that tangent unit vector [see Equation (13.2)].

$$V_h \equiv \overrightarrow{P_l P_c}, \tag{13.1}$$

$$\hat{V}_h = \frac{V_h}{|V_h|}, \tag{13.2}$$

where P_l is the previous position (x_l, y_l), P_c is the current position (x_c, y_c), and is the force direction unit vector. If the force value exceeded the maximum output (3.3 N), it was kept at this value.

With the increase in the transparency of the system, jitters generated from small delays and errors in the system often cause unacceptable oscillation of the tele-manipulator. The trade-off between system stability and transparency (matching level of the feedback forces and the forces applied at the tool tip) is a limitation of direct force feedback [16].

In order to validate the proposed experimental tele-manipulator and force feedback platform, a silicone phantom was fabricated (Figure 13.2). The motivation behind this approach is to develop a haptic feedback method that can be used by surgeons during robotic surgery and would assist detection and

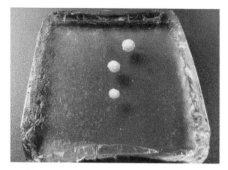

Figure 13.2   Soft tissue silicone phantom with embedded hard nodules to simulate tumors.

localization of stiffer malignant formations. Hard nodules simulating tumors of different sizes were embedded at different depths in the silicone phantom.

13.2.2 Evaluation of the Experimental Tele-manipulator and Force Feedback Platform

Ten human subjects (age range from 23 to 42) were asked to perform a palpation task with tele-operation system using force feedback. All tests were performed by human subjects controlling the slave robot to palpate a silicone phantom tissue with three nodules embedded through the stylus of the PHANToM Omni.

Sensitivity *Se* [17], which represents the test's ability to identify positive results, was defined as:

$$Se = \sum_{i=1}^{n} TP_i \Big/ \sum_{i=1}^{n} (TP_i + FN_i), \tag{13.3}$$

where *TP* is true positives and *FN* is false negatives.

Nodules could be detected using the experimental tele-manipulator and force feedback platform. The nodule detection sensitivity value was 76.7% (95% confidence interval: 59.1–88.2%).

13.3 Visual Stiffness Feedback

Alternative approach to standard haptic feedback methods via kinesthetic sensation is the use of visual modalities. In this section, we describe the use of visual stiffness feedback for RMIS. The benefit of such an approach in comparison to kinesthetic-based methods is that it can be applied in

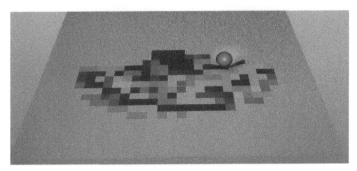

Figure 13.3   Stiffness distribution acquired from the surface of an organ.

combination with complex surgical manipulation when it is undesirable to disturb the performed manipulation. Alternatively, this method can be applied for the use on surgical simulator as a form of assessment. For instance, it is desirable to access the performance of training for a surgeon for a system that is not equipped with force and stiffness measurement devices.

Visual haptic feedback is used to represent the real-time distribution of stiffness acquired from the surface of an organ (Figure 13.3). RGB color-coded map is used to represent the relative stiffness of the tissue, thus, highlighting the areas of high stiffness that are likely to correspond to the location of tumors.

13.3.1 Experimental Setup to Validate the Concept of Visual Stiffness Feedback

To validate the feasibility of visual stiffness feedback, a tele-manipulation setup mimicking the arrangement of surgical robot for minimally invasive surgery was developed (see Figure 13.4). Fanuc robotic arm was used to represent surgical robot with a force sensitive probe for soft tissue palpation. Robotic arm (slave) was controlled by a user using commercially available haptic device (PHANToM Omni). Thus, a trajectory of a robot was defined and controlled by a user via PHANToM Omni device. As in this setting the target was to study visual stiffness feedback only, force feedback option of the haptic device was disabled, and it was used to control the robot position only. Robot trajectory was controlled in a tool coordinate space using position control. Control of a robot was implemented using Ethernet communication. To measure tissue stiffness, a probe with a spherical indenter and force and torque sensor was attached to the end effector of the robot.

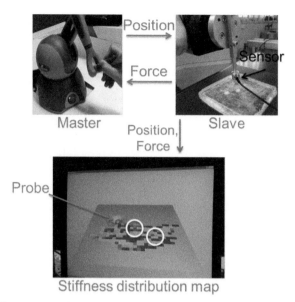

Figure 13.4 Components of tele-manipulation setup: Phantom Omni (master device) is used for position control; Fanuc robot arm (slave device) implements user-defined trajectory; force and torque sensor is embedded in the palpation tool and is used to measure tool and tissue phantom interaction forces; stiffness distribution map displays real-time results of palpation.

In order to validate the proposed approach, the same silicone phantom as described in the previous section was used. The motivation behind this approach is to develop a haptic feedback method that can be used by surgeons during robotic surgery and would assist detection and localization of stiffer malignant formations. The surface of the silicone was modeled on the screen and stiffness distribution was shown after palpation of the certain area. Subject performing remote palpation is facing screen with the simulated organ. Stiffness information was calculated in real time using force magnitude and displacement information of the tool of the robot. The characteristic feature of this algorithm is that the stiffness distribution is displayed as a relative value. An algorithm stores relative minimum and relative maximum values of stiffness. Further on, as soon as a harder region is encountered during palpation, the local maximum value is updated. Harder (stiffer) areas are displayed with higher intensity values of the RGB color map. This approach is feasible for real surgical application, as in most cases, alike during finger palpation, the knowledge about the value of stiffness is not required. More importantly, it is necessary to detect and localize areas of higher stiffness.

The limitation of this method is the need of a tissue model to create realistic stiffness distribution. However, an alternative approach is to employ three-dimensional visual information from endoscopic camera that is typically used together with surgical tools during keyhole surgery.

13.3.2 Evaluation of Visual Stiffness Feedback

Visual stiffness feedback was evaluated during experimental trials with human subjects. Participants were given a task to remotely palpate soft tissue phantom (Figure 13.2) and to explore the surface of a silicone organ using global palpation. Global palpation is a technique used to examine the whole surface of an organ and to define the areas that require thorough examination to determine the presence of abnormalities, such as tumors. This method is very useful to reduce the risk of missing any suspicious regions and can give general information about the state of an organ. As a standard palpation method, it is performed using fingers. In this section, we present the feasibility of the use of global palpation method for tele-manipulated surgery using visual stiffness feedback. There are several ways to perform global palpation technique depending on the type of organ [18, 19].

In the evaluation studies, subjects were first given freedom to apply any desired pattern, and then they were asked to execute the specific pattern of global palpation that is designed to improve palpation performance (Figure 13.7). The second pattern involved scanning of the organ surface from top to bottom using parallel paths, as well as applying circular palpation for the locations where abnormalities are suspected. In addition, circular patterns

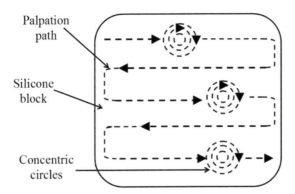

Figure 13.5 Schematic palpation trajectory pattern used to scan the whole surface of the target area with concentric circles applied in the areas of possible inclusions [20].

can give an indication of the stiffness and dimensions of an inclusion in case it is present. Ten subjects were given a task to identify the presence of three simulated abnormalities.

The results of experimental studies shown that subject can reliably use tele-manipulation setup to perform palpation and to identify the presence of hard abnormalities. Average performance of palpation for unrestricted pattern was 73%, while the use of prescribed pattern leads to a higher nodule detection rate with 90% of success. In addition, most subjects indicated that the use of prescribed pattern is easier to apply to detect hard nodules. The involved subjects had little or no palpation experience; however, no difficulties were reported in interpretation of haptic feedback. Therefore, this experiment demonstrates the feasibility of visual stiffness feedback for haptic information in minimally invasive surgery.

13.4 Pseudo-haptic Tissue Stiffness Feedback

Pseudo-haptic feedback combines visual feedback with the resistance of an isometric device [21]. Pseudoforce feedback is generated by changing the pressure on the isometric device [22]. For example, if the user is pressing a spring simulated by an isometric stick, the spring on the screen becomes shorter so that the user has an illusion that the stick is compressed by the user's hand. The stick itself is not compressed. Pseudo-haptic feedback was used to simulate several haptic properties [21], including friction [3], stiffness [3, 22], mass [23], texture [4], and force [24]. In this section, we describe the concept of pseudo-haptic tissue stiffness feedback.

13.4.1 The Concept of Pseudo-haptic Tissue Stiffness Feedback

Pseudo-haptic feedback technique is used to simulate tissue stiffness by changing the ratio of the speed of cursor movement to the speed of finger movement [5, 25]. The user then experiences a corresponding resistance when the cursor speed slows down. Tissue stiffness can also be estimated via visual feedback of the tissue deformation when the applied force is controlled. Visual feedback of tissue surface deformation is required to provide to the user a more realistic feedback of the tissue behavior during palpation. Non-homogeneous stiffness property of soft tissue can be expressed by integrating the sliding behavior simulated by pseudo-haptic feedback with visualization of soft tissue deformation (see Figure 13.6).

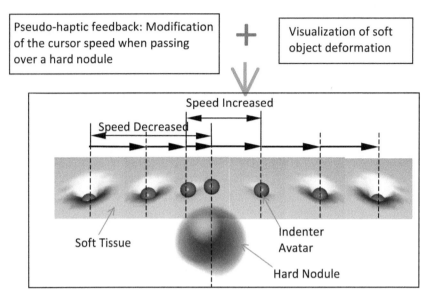

Figure 13.6 Combined pseudo-haptic feedback and visualization of tissue surface deformation [25].

13.4.2 The Combined Pseudo-haptic and Force Feedback

Since the mechanisms of pseudo-haptic feedback and force feedback are fundamentally different, they can be easily combined and will not adversely affect each other [26]. Force feedback was fed to the hand of the user through a haptic device, while the pseudo-haptic feedback information was fed via a graphical interface (see Figure 13.7). The force perception of the user was expected to come from a combination of sensations based on the proprioceptive and visual sensing of the user.

13.4.3 Evaluation of Pseudo-haptic Stiffness Feedback

An experimental validation study aiming at assessing the benefits of the proposed method was performed with the aim to define the efficiency of the proposed methods and to explore the advantages of using a combined pseudo-haptic and force feedback method.

Twenty participants were involved in the experiment: 6 women and 14 men (age range: 23–42). For each test, the same stiffness distribution was used, but the orientation of the silicone block or silicone block model was changed randomly from time to time. During the test, a stopwatch was used

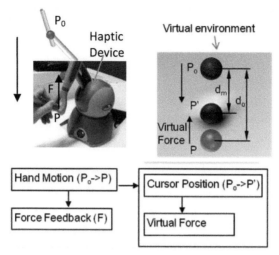

Figure 13.7 Combined pseudo-haptic and force feedback: the left panel is a haptic device, whose stylus is moved from P_o to P, and the right panel is a virtual environment, in which the cursor is supposed to move from P_o to P but actually moved to P' to create a virtual force.

in order to measure the time required by the participant to detect the hard nodules. The instrument allowed a precision of the time measurement of ±1 second.

For the test of pseudo-haptic feedback, participants were asked to do a practice run with visible hard nodule locations. Then, they were asked to look for hard nodules in the virtual soft object using pseudo-haptic feedback. The time taken to detect all nodules was recorded. For the test of the combination of pseudo-haptic feedback and force feedback, a practice run of the test was also first conducted. Then, participants were asked to examine the virtual soft object with embedded nodules by using the combined feedback to look for hard nodules. The time needed to detect all nodules was recorded for each participant. These two tests were conducted in a pseudo-random order.

The significance of the difference of sensitivity Se between the tests was examined. It was conducted by comparing the observed probabilities (p_1 and p_2) with a combined interval CI [27]:

$$CI = \sqrt{(P_1 - P_1)^2 + (P_2 - P_2)^2}, \tag{13.4}$$

where if $p_1 < p_2$, P_1 is the upper bound of p_1 and P_2 is the lower bound of p_2. If $|p_1 - p_2| > CI$, there is a significant difference between the two tests.

Wilson score intervals [28] were calculated at a 95% confidence level.

$$\frac{1}{1+\frac{z^2}{n}}\left[\hat{p}+\frac{z^2}{2n}\pm z\sqrt{\frac{\hat{p}(1-\hat{p})}{n}+\frac{z^2}{4n^2}}\right] \qquad (13.5)$$

where $\hat{p}$ is the proportion of successes estimated from the statistical sample; z is the $1-\alpha/2$ percentile of a standard normal distribution; α is the error percentile; and n is the sample size. Here, since the confidence level was 95%, the error α was 5%.

Positive predictive value (*PPV*) [29], which is a measure of the performance of the diagnostic method, was defined as:

$$PPV = \sum_{i=1}^{n} TP \Big/ \sum_{i=1}^{n} (TP + FP)$$

The technique using only pseudo-haptic feedback had a sensitivity *Se* of 50% (37.7–62.3%) overall while the combined technique utilizing both pseudo-haptic feedback and force feedback achieved a sensitivity *Se* of 83.3% with 95% confidential interval 71.9–90.7%. Compared to pseudo-haptic feedback, the *PPV* of the combination method was larger (94% versus 83.3%). Sensitivities *Se* and *PPV* were tested in difference significance. Table 13.1 shows the test results.

As shown in Figure 13.8, the combination method needed less time than the pseudo-haptic feedback test. A Wilcoxon signed-rank test was conducted

Table 13.1   Comparison of nodule detection sensitivities and positive predictive values

	CI	Δp	Significance?
Se	0.167	0.333	Yes
PPV	0.098	0.107	Yes

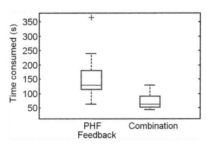

Figure 13.8   Time needed to find nodules using pseudo-haptic feedback and combination technique of pseudo-haptic feedback and force feedback.

Table 13.2 Wilcoxon signed-rank tests for consumed time

n_r	W	$W_{critical}$	Significance
19	4	46	$W < W_{critical}$, Yes

to test the significance of time difference. As shown in Table 13.2, the time difference was significant.

13.5 Haptic Feedback Actuators

Multi-fingered interaction is more common than single-fingered interaction in daily life, and is considered more efficient than single-fingered interaction when conveying haptic information [30]. While multi-fingered haptic feedback conveys more haptic information than single-point force feedback, the actuator elements in a multi-fingered force system can be much reduced compared to tactile haptic methods as, for example, described in [31] and [32]. There are some reports about multi-fingered force feedback for palpation [33–36]. Nevertheless, those devices are relatively expensive. In this section, we describe the use of fingertip haptic feedback actuators for RMIS.

13.5.1 Experimental Setup to Validate the Finger-tip Haptic Feedback Actuators

A pneumatic actuator was used to convey soft tissue stiffness information. Figure 13.10 shows the proposed pneumatic haptic feedback actuator. This actuator is consisted of a deformable surface [including a soft silicone layer (RTV6166 A:B = 1:2, thickness: 3 mm), a silicone rubber membrane (SILEX Ltd., HT6240, thickness: 0.25 mm, tensile strength: 11 N/mm^2, elongation at break: 440%, tear strength: 24 N/mm [37])], a non-deformable PDMS (poly-di-methyl-siloxane) substrate with a cylindrical hole (diameter of 4 mm), air

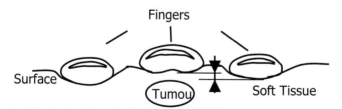

Figure 13.9 Multifinger interaction.

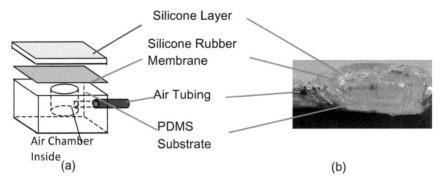

Silicone Layer

Silicone Rubber
Membrane

Air Tubing

PDMS
Substrate

Air Chamber
Inside

(a)

(b)

Figure 13.10   A pneumatic haptic feedback actuator: schematic diagram of the components shown in (a) and the prototype shown in (b) [38].

tubing, and a pressure-controllable air supply. When it is in use, air can be injected into the cavity and causes the silicone rubber membrane to inflate. The actuator creates a stress change on the user's fingertip and gives an impression of the stiffness change. The top soft silicone layer was applied to simulate the touch impression of soft tissue and limit the deformation of the silicone rubber film. Translucent silicone rubber adhesive E41 bounded the silicone rubber membrane and the substrate together. The air tubing was connected to the PDMS substrate by using RTV108 clear silicone rubber adhesive sealant. The PDMS substrate was fabricated using 3D prototyped molds.

Figure 13.11 illustrates the control of the proposed pneumatic haptic feedback actuator. The calculation of the three channels of air pressure values relates to the tactile sensing input (e.g., from a tele-manipulator). Data from the tactile sensor elements can be divided into three groups and the average values can be used as the input of a haptic feedback channel. Two NI USB-6211cards were used to generate analog signals for the pressure regulators (ITV0010, SMC). The air compressor (Compact 106, FIAC Air-Compressors) output was set to be 1500 kPa. The pressure regulators decreased the air pressure and inflated each of the actuators with proportional pressures to the analog signals ranging from 0 to 100 kPa. As shown in Figure 13.12, the proposed multi-fingered palpation can be used in RMIS.

To prove the efficiency of the proposed actuator, a user study on palpation using a premeasured stiffness distribution map was conducted. The experimental setup is shown in Figure 13.13. A pressure-sensitive touchpad (Wacom BAMBOO Pen & Touch) was used as an input device of position and normal force. The graphical feedback of the interaction on the tissue

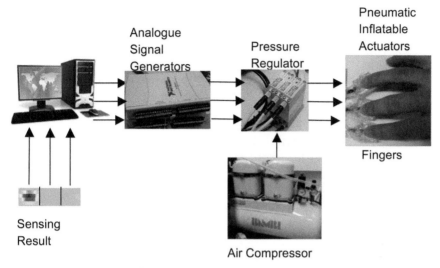

Figure 13.11   Multifingered palpation system [38].

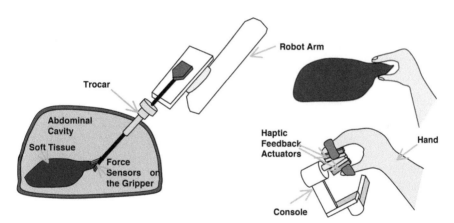

Figure 13.12   Schematic diagram of the applications of the proposed multi-fingered palpation in robot-assisted minimally invasive surgery (RMIS).

surface through computer graphics and the multi-fingered haptic feedback were provided. Three colored spheres displayed on the graphical interface were used to represent three fingers. These three spheres were aligned in a right-angled triangular shape and were set to follow the motion of the pen. The output forces via the pneumatic actuators to the three fingers were translated independently from each other according to the applied palpation

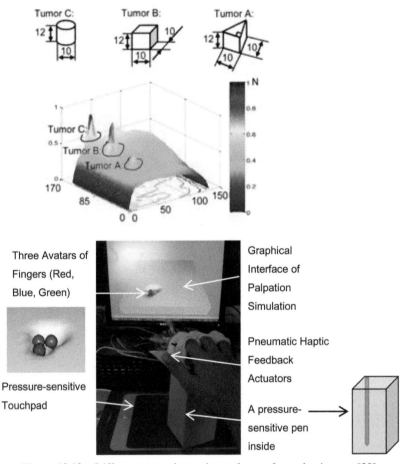

Figure 13.13   Stiffness map and experimental setup for evaluation test [38].

force on the touchpad and the local stiffness value. In this way, users were able to explore three neighboring properties simultaneously.

13.5.2 Evaluation Results of Finger-tip Haptic Feedback Actuators

Nine subjects were involved in this study (age range: 23–36, ♀: 4; ♂: 8). None of them had any palpation experience. In the experiment, all participants could feel the simulated stiffness differences. The measured stiffness distribution came from a silicone phantom soft tissue embedded with artificial tumors A, B, and C (see Figure 13.13). Tumors were represented as plastic cubes with thicknesses of 4, 12 and 8 mm. The detection sensitivities *Se* of simulated tumor A, B, and C were 66.7, 100, and 88.9%, respectively. There was a positive correlation between the nodule detection sensitivities and nodule sizes.

13.6 Conclusion

In this chapter, haptic feedback modalities for minimally invasive surgery to compensate the loss of physical contact between the surgeon and the soft tissue were introduced and validated, including force feedback from an experimental tele-manipulator and the force feedback platform, visual stiffness feedback representing stiffness distribution of soft tissue, pseudo-haptic feedback expressing haptic perception through a visual display, and finger-tip haptic feedback actuators enabling multi-fingered haptic feedback. Those proposed methods can also be adopted for other applications where sensory substitution is required, including VR-based games and general robotic manipulation.

Acknowledgments

The authors thank the participants of the experiments for their contributions.

References

[1] Gwilliam, J. C., Pezzementi, Z., Jantho, E., Okamura, A. M., and Hsiao, S. (2010)."Human vs. robotic tactile sensing: detecting lumps in soft tissue," in *Proceedings of the IEEE Haptics Symposium*, 21–28.

[2] Tavakoli, M., Aziminejad, A., Patel, R. V., and Moallem, M. (2006). Methods and mechanisms for contact feedback in a robot-assisted minimally invasive environment. *Surg. Endosc.* 20, 1570–1579.

[3] Lécuyer, A., Coquillart, S., Kheddar, A., Richard, P., and Coiffet, P. (2000). "Pseudo-haptic feedback: can isometric input devices simulate force feedback?," in *Proceedings of the IEEE Virtual Reality*, New Brunswick, NJ, 83–90.

[4] Lécuyer, A., Burkhardt, J.-M. and Tan, C.-H. (2008). A study of the modification of the speed and size of the cursor for simulating pseudo-haptic bumps and holes. *ACM Trans. Appl. Percept.* 5, 1–21.

[5] Li, M., Liu, H., Li, J., Seneviratne, L. D., and Althoefer, K. (2012). "Tissue stiffness simulation and abnormality localization using pseudo-haptic feedback," in *Proceedings of the IEEE International Conference on Robotics and Automation*, Saint Paul, MN.

[6] Trejos, A. L., Jayender, J., Perri, M. T., Naish, M. D., Patel, R. V., and Malthaner, R. A. (2008). "Experimental evaluation of robot-assisted tactile sensing for minimally invasive surgery," in *Proceedings of the 2nd IEEE RAS EMBS International Conference on Biomedical Robotics and Biomechatronics*, Scottsdale, AZ, 971–976.

[7] Perri M. T., and Trejos, A. L. (2010). New tactile sensing system for minimally invasive surgical tumour localization. *Int. J. Med. Robot. Comput. Assist. Surg.* 6, 211–220.

[8] Perri, M., Trejos, A., and Naish, M. (2010). "Initial evaluation of a tactile/kinesthetic force feedback system for minimally invasive tumor localization," in *Proceedings of theIEEE/ASME Transactions on Mechatronics*, 15, 925–931.

[9] Miller, A. P., Peine, W. J., Son, J. S., and Hammoud, M. D. Z. T. (2007). "Tactile imaging system for localizing lung nodules during video assisted thoracoscopic surgery 1," in *Proceedings of the IEEE International Conference on Robotics and Automation*, Roma, 10–14.

[10] King, C.-H., Culjat, M. O., Franco, M. L., Bisley, J. W., Carman, G. P. Dutson, E. P., et al. (2009). "A multielement tactile feedback system for robot-assisted minimally invasive surgery," in *Proceedings of the IEEE Transactions on Haptics*, 2, 52–56.

[11] Liu, H., Noonan, D. P., Challacombe, B. J., Dasgupta, P., Seneviratne, L. D., and Althoefer, K. (2010). "Rolling mechanical imaging for tissue abnormality localization during minimally invasive surgery," in *Proceedings of the IEEE Transactions on Biomedical Engineering*, 57, 404–414.

[12] Liu, H., Li, J., Poon, Q., Seneviratne, L. D., and Althoefer, K. (2010). "Miniaturized force-indentation depth sensor for tissue abnormality identification during laparoscopic surgery," in *Proceedings of the International Conference on Robotics and Automation (ICRA)*, Anchorage, AK, 3654–3659.

[13] Noonan, D. P., Liu, H., Zweiri, Y. H., Althoefer, K. A., and Seneviratne, L. D. (2007). "A dual-function wheeled probe for tissue viscoelastic property identification during minimally invasive surgery," in *Proceedings of the IEEE International Conference on Robotics and Automation*, Roma, 2629–2634.

[14] Zirjakova, J. (2011). *Prostate Post-surgical 3D Imaging and Data Analysis*. London: King's College.

[15] Liu, H., Li, J., Song, X., Seneviratne, L. D., and Althoefer, K. (2011). Rolling indentation probe for tissue abnormality identification during minimally invasive surgery. *IEEE Trans. Robot.* 27, 450–460.

[16] Okamura, A. M. (2009). Haptic feedback in robot-assisted minimally invasive surgery. *Curr. Opin. Urol.* 19, 102–107.

[17] Altman, D., and Bland, J. (1994). Diagnostic test 1: sensitivity and specificity. *BMJ.* 308:1552.

[18] Wang, N., Gerling, G. J., Childress, R. M., and Martin, M. L. (2010). "Quantifying palpation techniques in relation to performance in a clinical prostate exam," in *Proceedings of the IEEE Transactions on Information Technology in Biomedicine* 14, 1088–1097.

[19] Saunders, K. J., Pilgrim, C. A., and Pennypacker, H. S. (1986). Increased proficiency of search in breast self-examination. *Cancer* 58, 2531–2537.

[20] Konstantinova, J., Li, M., Aminzadeh, V., Althoefer, K., Dasgupta, P., and Nanayakkara, T. (2013). "Evaluating manual palpation trajectory patterns in tele-manipulation for soft tissue examination," in *Proceedings of the IEEE International Conference on Systems, Man, and Cybernetics*, 4190–4195.

[21] Lécuyer, A. (2009). Simulating haptic feedback using vision: a survey of research and applications of pseudo-haptic feedback. *Presence Teleoperators Virtual Environ.* 18, 39–53.

[22] Lécuyer, A. (2001). "Simulating haptic information with haptic illusions in virtual environments," in *Proceedings of the NATO RTA/Human Factors and Medicine Panel Workshop*.

[23] Dominjon, L., Lecuyer, A., Burkhardt, J., Richard, P., and Richir, S. (2005). "Influence of control/display ratio on the perception of mass of manipulated objects in virtual environments," in *Proceedings of the IEEE Virtual Reality*, Bonn, 19–25.

[24] Pusch, A., Martin, O., and Coquillart, S. (2009). HEMP—hand-displacement-based pseudo-haptics: A study of a force field application and a behavioural analysis. *Int. J. Hum. Comput. Stud.* 67, 256–268.

[25] Li, M., Sareh, S., Ridzuan, M., Seneviratne, L. D., Dasgupta, P., Wurdemann, H. A., and Althoefer, K. (2014). "Multi-fingered palpation using pseudo-haptic feedback," in *Proceedings of the The Hamlyn Symposium on Medical Robotics*, London, 3, 8–9.

[26] Hachisu, T., Cirio, G., Marchal, M., and Lécuyer, A. (2011). "Pseudo-haptic feedback augmented with visual and tactile vibrations," in

Proceedings of the IEEE International Symposium on Virtual Reality Innovation, Singapore, 1, 327–328.

[27] Wallis, S. (2013). Binomial confidence intervals and contingency tests: mathematical fundamentals and the evaluation of alternative methods. *J. Quant. Linguist.* 20, 178–208.

[28] Wilson, E. B. (1927). Probable inference, the law of succession, and statistical inference. *J. Am. Stat. Assoc.* 22, 209–212.

[29] Fawcett, T. (2006). An introduction to ROC analysis. *Pattern Recognit. Lett.* 27, 861–874.

[30] Dinsmore, M., Langrana, N., Burdea, G., Ladeji, J., and Box, P. O.(1997). "Virtual Reality Training Simulation for Palpation of Subsurface Tumors," in *Proceedings of the IEEE Virtual Reality Annual International Symposium*, Albuquerque, NM, 54–60.

[31] Kim, S.-Y., Kyung, K.-U., Park, J., and Kwon, D.-S. (2007). Real-time area-based haptic rendering and the augmented tactile display device for a palpation simulator. *Adv. Robot.* 21, 961–981.

[32] Culjat, M., King, C.-H., Franco, M., Bisley, J., Grundfest, W., and Dutson, E. (2008). Pneumatic balloon actuators for tactile feedback in robotic surgery. *Ind. Robot An Int. J.* 35, 449–455.

[33] Langrana, N. A., Burdea, G., Lange, K., Gomez, D., and Deshpande, S. (1994). Dynamic force feedback in a virtual knee palpation. *Artif. Intell. Med.* 6, 321–333.

[34] Daniulaitis, V., and Alhalabi, M. O. (2004). "Medical palpation of deformable tissue using physics-based model for haptic interface robot (HIRO)," in *Proceedings of the 2004 IEEE/RSJ International Conference on Intelligent Robots and Systems,* Sendai, 3907–3911.

[35] Kawasaki, H., Takai, J., Tanaka, Y., Mrad, C., and Mouri, T. (2003). "Control of multi-fingered haptic interface opposite to human hand," in *Proceedings of the 2003 IEEERSJ International Conference on Intelligent Robots and Systems IROS 2003*, Las Vegas, NV, 3, 2707–2712.

[36] Endo, T., Kawasaki, H., Mouri, T., Doi, Y., Yoshida, T., Ishigure, Y., Shimomura, H., Matsumura, M., and Koketsu, K. (2009). "Five-fingered haptic interface robot: HIRO III," in *Proceedings of the World Haptics 2009 Third Joint EuroHaptics conference and Symposium on Haptic Interfaces for Virtual Environment and Teleoperator Systems*, 4, 458–463.

[37] S. S. Ltd. (2014). *"HT 6240 data sheet."*

[38] Li, M., Luo, S., Nanayakkara, T., Seneviratne, L. D., Dasgupta, P., and Althoefer, K. (2014). Multi-fingered haptic palpation using pneumatic feedback actuators. *Sensors Actuators A Phys.* 218, 132–141.

[39] Yamamoto, T., and Abolhassani, N. (2012). Augmented reality and haptic interfaces for robot-assisted surgery. *Int. J. Med. Robot. Comput. Assist. Surg.* 8, 45–56.

14

Force Feedback Sleeve Using Pneumatic and Micro Vibration Actuators

Łukasz Mucha and Krzysztof Lis

Prof. Z. Religa Foundation of Cardiac Surgery Development, Zabrze, Poland

Abstract

The search for the best solution from the point of view of achieving a construction imitating the contact between the operator of the surgical device and the obstacle resulted in the origin of a concept involving the use of a special armband or a sleeve.

This chapter presents two types of haptic feedback concepts that can be used to translate force feedback from the sensors of STIFF-FLOP system. The first concept is based on the design concept of a pneumatic feedback system, and on the working principles of a blood pressure measuring sleeve used in medical diagnostics.

The second concept includes the use of miniature seismic inductors. In addition, this chapter reviews a design of various inductors. This interesting area of human–machine interface can significantly increase the amount of information coming to the operator allowing for more precise and safe control. This solution may help to reduce the risk of operation with the robot in the surgical robot control console.

14.1 Introduction

When discussing human–machine communication, force feedback is one of the main discussed issues. It is believed that feedback is one of the essential elements needed in order for humans to control machines effectively. Feedback can simply be defined as providing the operator with information about the results of his/her work. Development of feedback defined in such a way

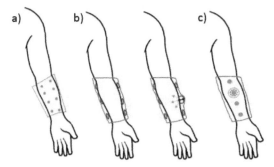

Figure 14.1   The concept of an armband for surgeon's forearm (a) arrangement; (b) concept of pneumatic airbags; and (c) application of electric motors.

focuses on increasing the amount and the quality of the information transfer between the device and its operator [1–3]. Introducing tactile feedback during surgical procedures performed by surgical robots can help surgeons to sense the characteristics of specific tissues, recognizing pathological states or applying precise surgical suture tension. The application of feedback in robotic surgery can also have a positive effect on the learning curve associated with robot operation [4–8]. The search for the best solution from the point of view of achieving the construction that can imitate the contact between the operator of the surgical device and the obstacle resulted in the origin of a concept involving the use of a special armband (sleeve). This armband is placed by the surgeon on his/her forearm and it provides additional information from the operating field (Figure 14.1). Between the skin and the armband, there are mechanisms which can produce tactile stimuli. Mechanic solutions based on pneumatic and electric vibration motors were chosen as a mean of the interaction factor.

14.2 Application of the Pneumatic Impact Interaction

The STIFF-FLOP project involved the construction of pneumatic actuators along the forearm and around it. Figure 14.2 presents the 4×5 matrix of 55×30-mm pneumatic actuators.

The design concept of a pneumatic feedback system is based on the working principles of a blood pressure measuring sleeve used in medical diagnostics. The elastic airbags are made of two layers of vulcanized (or adhesive) rubber (see Figure 14.3). Before the process of vulcanization, plastic tubes are placed between the layers in such a way that each tube is

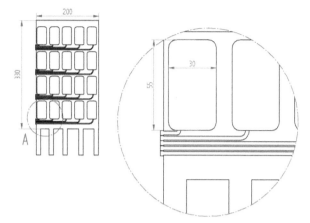

Figure 14.2   Design concept of the matrix of airbags' dimensions.

Figure 14.3   Design concept of the pneumatic sleeve [9].

able to feed the airbag created in the process of vulcanization. The tubes exit through the packets feeding the airbags along the forearm. The airbags are covered with a layer of elastic fabric from the side of the contact with the skin of the hand and with a more stiff fabric on the outer side which prevents the pressure exerted by the airbags working in the opposite direction than desired (hand). Fixing the sleeve on a hand and adjusting it to the individual characteristics of the operator can be done with the help of Velcro straps.

14.3 Control

In order to make things easier for the operator, the control unit and power supply module were removed from the sleeve. Only the airbags and pneumatic leads were left. The first concept involved controlling the pneumatic sleeve in a continuous manner. The value of airbag's impact on the operator

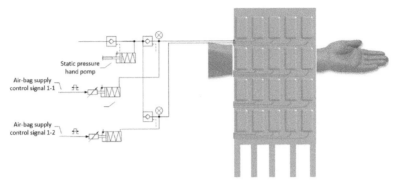

Figure 14.4   Concept of the continuously sleeve control system [9].

depended on the current flowing in the coil of the electro-pneumatic transducer.

The coil's pulse-width modulation (PWM) signal controlled the pressure generated by the transducer. This concept does not require compressed air supply, and the pneumatic system does not generate any noise. Figure 14.4 presents the concept of controlling two airbags. Eventually, it is replicated (4 × 5) times.

The second type of sleeve control was discrete. The operator could feel or not the impact of the airbag – the so-called two-state control. After fixing the sleeve with the Velcro straps, the airbags were pumped so that the sleeve is in close contact with the skin of the operator but would not cause too much pressure and the resulting discomfort (Figure 14.5a).

During evaluation tests (Figure 14.6), it was observed that the second type of control was more easily perceptible to the operator. However, it was later modified to the form presented in the drawing in Figure 14.5b. Simplification of the pneumatic system resulted in more favorable subjective perceptions experienced by the testing group.

Loud operation of the pneumatic system, the elaborate control system, the necessity of providing each individual airbag with a lead, and a limited movement of the operator required a change of concept. Therefore, the next version of the sleeve was built using the electro-mechanical vibration motors.

14.4 Applications of Electric Vibration Motors

Another concept involved equipping the sleeve with micro seismic vibration motors as devices mechanically interacting with a human. This solution

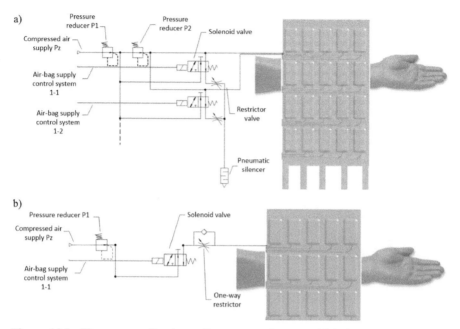

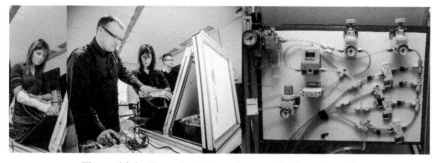

Figure 14.5   The concept of a sleeve discrete control system (a) before and (b) after the modifications.

Figure 14.6   Operation tests andpneumatic sleeve control.

includes fixing some vibration motors on the operator's forearm with the purpose of mechanically signaling the interactions in the operating field. The motors are arranged as a 4 × 4 matrix in the sleeve worn by the operator during his/her work. Choosing the right construction of the motor is a difficult task as the offered micro motors differ in size, mass, manner of work, and generated power. Therefore, the solution is a compromise between mass and size (the smaller the better) and generated power (the more the better) [10, 11]. The solutions of the most popular constructions of mechanical micro motors are shown in Figure 14.8. The principle of operation of the first one is based on the VCM (voice coil motor) (Figures 14.7a and b). It operates by generating vibrations through seismic mass set in reciprocating motion by means of electromotive force. Figures14.7c–f show motors generating vibrations through eccentric mass mounted on the rotor of a DC motor.

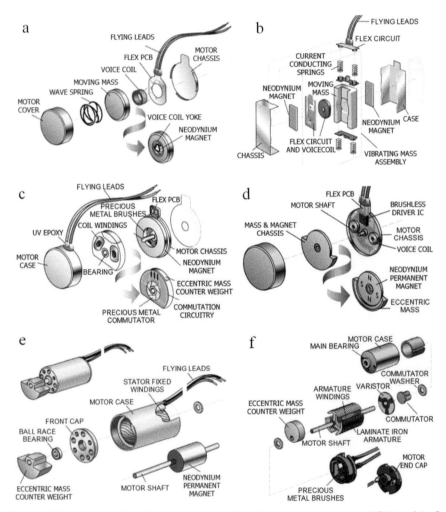

Figure 14.7 Construction of micro motors: (a, b) a linear resonant actuator (LRA) and (c–f) an eccentric rotating mass vibration motor (ERM) [12].

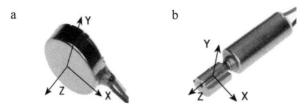

Figure 14.8   Micro motors with DC motors: (a) coin motor and (b) cylindrical motor.

Motors with vibrating mass can have two types of constructions, the first of which is associated with motor type and the other with dimension variant. Figure 14.7f presents a motor built on a conventional DC motor and Figures 14.7d and e based on a brushless motor. The main distinguishing feature of these structures is their control and durability. Brushless motors are cheaper and easier to control but they are less durable. The operating parameters of the two solutions are similar [12].

Other division of the dimension variant involves splitting the motors into coin motors (Figures 14.7a–d) and cylindrical motors (Figures 14.7e and f). The main difference is associated with the motor case. In this regard, the coin motor (shown in Figure 14.8a) is more favorable because it does not require additional safety casing.

The functionality of the sleeve dictates the mounting method of the selected motors, so for the coin motor (Figure 14.8a) the XY surface should be parallel to the body surface. The cylindrical motor (Figure 14.8b), on the other hand, should be mounted on the ZX or ZY surface parallel to the body surface. The sleeve design uses the coin type. This motor generates vibrations in the YX plane (Figure 14.8a).

The sleeve prototype was made in two versions. Both versions were made of elastic fabric. The first version (Figure 14.9) was too rigid and limited operator's range of motion. Moreover, it turned out that vibrations were less perceptible compared to the second version.

The second version of the sleeve was based on the same motors. However, other fabrics (polyamide and elastane) were used to improve the ergonomics when using the sleeve and increase the perceptibility to a certain degree (Figure 14.10). The control system was also modified by introducing a gradation of vibration intensity.

The first prototypes were equipped with a PCB placed in the sleeve and powered by an external power supply and connected to a computer with an RS 232 cable. Next, an application for computers was made using the RAD

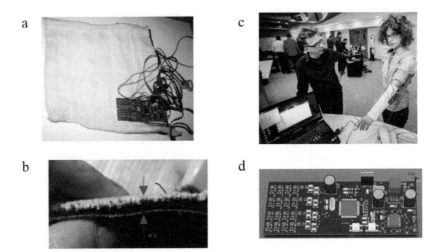

Figure 14.9   Sleeve design version 1: (a) sleeve; (b) crosssection of layers of fabric; (c) test bench; and (d) control board based on thecortex F4 microcontroller.

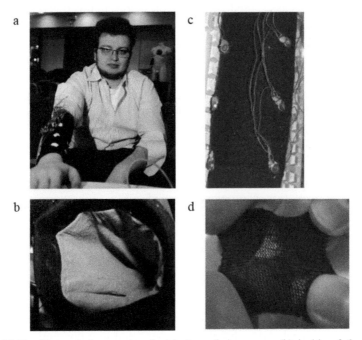

Figure 14.10   Sleeve design version 2: (a) sleeve being worn; (b) inside of the sleeve; (c) arrangement of motors; and (d) material used in the outer layer.

Studio software that allows the user to set different functions. In the first version, the sleeve motors' control system was discrete (on/off). In the second version of the sleeve, the control was based on the PWM signal.

Changing the performance of the motor is difficult and possible only to a small extent. Therefore, the experimental value of PWM control signal frequency with subcarrier was set to approximately 2 Hz. This value is closely related to the start and stop characteristics of the motor. Accepted discrete values are as follows:

- 100% defined as "strong"
- 50% defined as "average"
- 20% defined as "weak."

Subjective tests using the above values were conducted with people. Tests on the subjective perception of the location and the intensity of the sensation caused by a single motor were run in a group of seven people. During the tests, the user was positioned in the same position as the operator of the surgical robot console during the operation with arms outstretched and supported at the elbows. The purpose of the test was for the group to determine the most user-friendly and most perceptible (effective) motor-control systems. In order to achieve that, three types of motor signals were set: weak, average (pulsating), and strong. The motors were switched individually in a random order.

During the test, the operator was asked to indicate the number of the motor switched on and to determine the vibration intensity. After indicating the correct number of the working motor and the vibration intensity, the answer was accepted.

The motors were placed on the operator's hand in such a way as to maintain maximum distance between the neighboring motors (Figure 14.11).

The preferred control signal was a high-amplitude, continuous signal. In this case, the average accuracy of indicating the vibrating motor was 95%. The sensitivity matrix can be seen in Figure 14.12.

14.5 Conclusion

After the initial testing of the pneumatic sleeve and the modifications of the control system, it was decided to change the concept.

Further work was carried out using the electro-mechanical vibration motors.

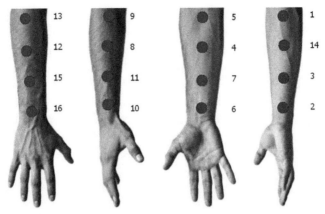

Figure 14.11   View of the motors' arrangement on the operator's forearm [13].

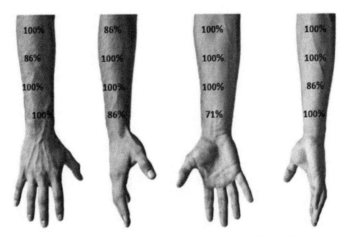

Figure 14.12   Test results – operator's subjective response to the third control signal [13].

A 16-vibrator matrix deployed in four cross-sections and four longitudinal sections of the sleeve has been placed in the sleeve to simulate the feeling of interaction between the hand and the environment. In the first prototype of the sleeve, it was difficult to indicate the place of the vibration due to the stiffness of the fabric used. The steering of a single vibrator was realized in a discrete manner – the vibrator was either on or off. Further research provided the information necessary for building the next version. Changes made in the next version of the prototype allowed us to achieve three levels of intensity of the sensation. The accuracy of the subjective identification of the area of vibration by the testing group was over 80% in general, while in the case of the medium and strong intensity level, it was over 95%.

Based on the results of the tests, it can be concluded that the use of electro-mechanical vibration motors with an acceleration value of about 1 g and a frequency in the range of 100–200 Hz might be an innovative way of gaining device-operator feedback. It is possible to identify the location of interaction but it depends mostly on the physiological characteristics of the operator [13]. This interesting area of human–machine interface may significantly increase the amount of information reaching the operator and thus allow for more precise and safe control. The use of this solution in a surgical robot control console may help to reduce the risks of operating with the usage of robots.

Acknowledgments

The project of flexible tool was supported in part by the European Commission within the STIFF-FLOP FP7 European project FP7/ICT-2011-7-287728. The authors thank K. Rohr and Z. Nawrat from the Prof. Z. Religa Foundation of Cardiac Surgery Development for their assistance with the technical aspects of the study.

References

[1] Nadrag, P., Temzi, L., Arioui, H., and Hoppenot, P. (2011). "Remote control of an assistive robot using force feedback," in *Proceedings of the 15th IEEE International Conference of Advanced Robotics*, Tallinn, Estonia, 20–23.

[2] Liu, Y. K., Zhang, Y. M., Fu, B., and Yang, R. G. (2013). "Predictive control for robot arm teleoperation," in *Proceedings of the 39th Annual Conference on IEEE Industrial Electronics Society*, 3693–3698.

[3] Komada, S., and Ohnishi, K. (1990). Force feedback control of robot manipulator by the acceleration tracing orientation method. *IEEE Trans. Ind. Electron.* 37, 6–12.

[4] Maeda, S., Tsujiuchi, N., Koizumi, T., Sugiura, M., and Kojima, H. (2011). "Development and control of pneumatic robot arm for industrial fields," in *Proceedings of the 37th IEEE IECON/IECON*, 86–91.

[5] Uddin, M. W., Zhang, X., and Wang, D. (2016). "A pneumatic-driven haptic glove with force and tactile feedback," in *Proceedings of the 2016*

International Conference on Virtual Reality and Visualization (ICVRV), Hangzhou, 304–311.

[6] Du, H., Xiong, W., Wang, Z., and Chen, L. (2011). "Design of a new type of pneumatic force feedback data glove," in *Proceedings of the 2011 International Conference on Fluid Power and Mechatronics*, Beijing, 292–296.

[7] Gerboni, G., Diodato, A., Ciuti, G., Cianchetti, M., and Menciassi, A. (2017). "Feedback control of soft robot actuators via commercial flex bend sensors," in *Proceedings of the IEEE/ASME Transactions on Mechatronics*, 22, 4, 1881–1888.

[8] Culjat, M., King, C., Franco, M., Bisley, J., Grundfest, W., and Dutson, E. (2008). Pneumatic balloon actuators for tactile feedback in robotic surgery. *Ind. Robot: Int. J.* 35, 449–455.

[9] Wurdemann, H. A., Secco, E. L., Nanayakkara, T., Althoefer, K., Lis, K., Mucha, L., et al. (2013). "Mapping tactile information of a soft manipulator to a haptic sleeve in RMIS," in *Proceedings of the 3rd Joint Workshop on New Technologies for Computer Robot Assisted Surgery*, Verona.

[10] Wurdemann, H. A., Jiang, A., Nanayakkara, T., Seneviratne, L. D., and Althoefer, K. (2012). "Variable stiffness controllable and learnable manipulator for MIS," in *Proceedings of the Workshop: Modular Surgical Robotics: How Can We Make it Possible? IEEE International Conference on Robotics and Automation (ICRA)*.

[11] Maereg, A. T., Nagar, A., Reid, D., and Secco, E. L. (2017). Wearable vibrotactile haptic device for stiffness. Discrimination during virtual interactions. *Front. Robot.* 4:42.

[12] Precision Microdrives Ltd., Unit 1.05, London, United Kingdom. Available at: http://www.precisionmicrodrives.com.

[13] Mucha, Ł., Lis, K., and Rohr, K. (2013). "Wykorzystanie mikrowzbudników drgań do realizacji siłowego sprzężenia zwrotnego w zadajniku," in *Proceedings of the Postępy Robotyki Medycznej*, Rzeszów.

15

Representation of Distributed Haptic Feedback Given via Vibro-tactile Actuator Arrays

Anuradha Ranasinghe[1], Ashraf Weheliye[2], Prokar Dasgupta[3], and Thrishantha Nanayakkara[4]

[1]Department of Mathematics and Computer Science, Liverpool Hope University, United Kingdom
[2]Department of Informatics, King's College London, United Kingdom
[3]MRC Centre for Transplantation, DTIMB & NIHR BRC, King's College London, United Kingdom
[4]Dyson School of Design Engineering, Imperial College London, United Kingdom

Abstract

There are many studies to suggest that the human motor system adaptively combines motor primitives to control limb movements. However, little is known about whether the somatosensory system too uses a similar strategy to efficiently represent haptic experiences. Since it has been shown that humans learn movements through flexible combination of primitives that can be modeled using Gaussian-like functions, we try to explore whether human brain uses similar primitives to represent haptic memory too. We tested whether haptic memory is localized and magnitude specific along the arm. Therefore, experiments were conducted to understand how humans trained in primitive haptic patterns given using a wearable sleeve can recognize their shifts and linear combinations. A wearable sleeve was used that consisted of seven vibro-actuators to convey the primitive patterns. We found that (1) subjects find it easier to recognize uni-modal haptic templates. When they are given bi-modal patterns, subjects tend to generalize them to uni-modal patterns; (2) haptic memory is location specific. When the same template is shifted along the arm, the original template interferes with the shifted pattern. (3) subjects can recognize linear combinations of previously trained haptic templates. In addition to the prototype presented in Chapter 14, this chapter provides guidelines to design haptic feedback system for STIFF-FLOP continuum manipulator. The scenarios can include separation of force and torque into different distributed haptic feedback templates, and recognition of tissue contact forces at multiple points along the arm using optimally designed feedback templates using vibro-tactile actuator arrays.

15.1 Introduction

Haptics would be the best way to convey messages in critical tasks to provide spatial information where vision and audition are less reliable [1]. Therefore, as a first step, we try to understand how distributed haptic feedback is generalized at the somatosensory system. In human motor control studies, it has been shown that humans learn movements through a flexible combination of primitives that can be modeled using Gaussian-like functions [2]. How the somatosensory system represents distributed haptic patterns is not well known. In this paper, we address the question as to how a trained Gaussian template is recognized when it is scaled and shifted along the arm. The used primitive haptic patterns are the Gaussian template (T), T shifted right (TR), T shifted left (TL), half amplitude of T (THA), and half standard deviation of T (THS) hereafter referred to be templates. This approach will inform us to design new training protocols of a new device to provide haptic feedback.

There have been some studies demonstrating that haptic feedback can be used to assist humans [3–5]. Moreover, there have been many studies on using vibro-actuators for different purposes to convey messages to humans. For example, the study in [5] presented an active belt which is a wearable tactile display that can transmit multiple directional information in combination with a global positioning system (GPS) directional sensor and Seven vibro-actuators.

Another research on cooperative human robot haptic navigation in [6] used a wrist belt with vibro-tactile sensors to guide a human to a target location recently. Moreover, haptic feedback was used to navigate people by using a mobile phone in [7]. Furthermore, vibro-tactile way-point navigation was presented in [8] in pedestrian navigation. Vibration stimulation was used to convey different types of messages to the users in guiding with and without vision in [5] and [7], respectively. Mobile devices were used to provide pedestrian navigation systems in low visibility conditions [8–10].

In some studies, vibro-tactile displays have used to improve the quality of life in different ways such as reading devices for those with visual impairments [11] to provide feedback of body tilt [12], balance control and postural stability [13], and navigation aid in unfamiliar environments [14]. Until now, there have been many studies that used vibro-actuator belts for different purposes. However, this paper attempts to understand how to use vibro-tactile actuator arrays to understand representation of distributed haptic feedback.

Different studies introduced various design criteria of the vibro-actuator arrays depending on the objectives/hypothesis of the tests. For example, considering the location of vibro-actuators, it is noted that the accuracy was only 53% when subjects were asked to identify the location of vibro-tactile stimulation on the skin in [15]. Since the localization accuracy is low as noticed in [15], it is important to study how humans would generalize haptic feedback patterns to a given set of discrete locations of the skin, because an attempt to generalize may change the level of error at individual locations. Moreover, the two-point resolution varies across the body according to the study in [16]. For example, two-point resolution for the forearm is more than 35 mm [16]. Therefore, the gap between the vibro-actuators was maintained more than 35 mm to avoid unnecessary ambidextrous crosstalks in our experiments.

The experimental results in [17] show that the accuracy of tactile perception significantly improves when the duration of tactile stimulation lies in the range of 80–320 ms. Studies on stimulation time period in [18] showed that people prefer that the stimulus is between 50 and 200 ms and longer durations are perceived to be annoying. The study in [16] experienced that shorter length of vibrating time reduces the level of adaptation. Therefore, the durations of these experimental trials are limited to 80 ms.

Amplitude has been widely used to stimulate the human skin in most of the previous studies [19–22]. However, we argue that frequency would be better for persistent perception. The monotonic nature in amplitude could affect humans' responses. For example, humans feel more comfortable to clocks tick-tock than a hump because of the discrete nature of the former. The frequency was chosen in the most power-efficient region of the vibro-actuator.

Apart from previous studies on using vibro-actuator arrays mainly in navigation and health care, they were widely used to understand how somatosensory receptors work on haptic pattern recognition [23–25]. The ability of humans to identify patterns of vibro-tactile stimulation was tested using tactile displays mounted on the arm and torso [23]. The results in [23] indicated that the identification of the vibro-tactile patterns was superior on the torso compared to the forearm, with participants achieving higher accuracy with seven of the eight presented patterns. Haptic recognition of familiar objects by the early blind, the late blind, and the sighted was investigated with two-dimensional (2-D) and three-dimensional (3-D) stimuli produced by small tactor-pins in [24]. The results show that the haptic legibility of the 3-D stimuli was significantly higher than that of the 2-D stimuli for all the groups suggesting that 3-D presentation seems to promise a way to overcome the limitation of 2-D graphic display. Moreover, participants were trained to varying degrees of accuracy on tactile identification of two-dimensional patterns in [25]. Recognition of these patterns, of inverted versions of these patterns, and of subparts of these patterns was then tested in [25]. Those results in previous studies show the humans' ability to recognize different haptic feedback patterns in body location, and 2-D or 3-D when the stimulation was given via vibro-tactile arrays.

The scientific questions tested in this paper are how the human somatosensory system encodes Gaussian-like template haptic patterns and whether these representations are localized (cannot be shifted along the skin) and magnitude specific (cannot be scaled). Moreover, we ask whether humans can recognize two humped haptic patterns. The results would show how haptic perception is represented in the somatosensory neural circuits. i.e., how localized haptic memories are. These insights will helps to develop more efficient haptic feedback systems using a small number of templates to be learnt to encode complex haptic messages.

Therefore, to test those questions, the experiments were carried out to study humans' ability (1) to generalize (scaling/shifting) the trained primitive vibro-actuator array templates [Gaussian template (T), T shifted right (TR), T shifted left (TL), half amplitude of T (THA), and half standard deviation of T (THS)], (2) to recognize trained these templates and their inverse even they played randomly, and (3) to recognize these trained templates linear combinations of those patterns given via a vibro-actuator array. Therefore, this is the first paper that attempts to show how the primitive haptic-based patterns are represented by humans. The results of this paper would give an idea as to how humans construct the cutaneous feedback in different messages/scenarios. The results would help us to understand what the sensitive geometrical shapes are when we need to code haptic messages to humans in noisy crowded areas such as factory, search, and rescue via cutaneous feedback.

15.2 Materials and Methods

The results of three experiments would answer the following scientific questions: experiment 1: How humans generalize a Gaussian template in scaling and shifting, experiment 2: How humans can recognize trained templates and their inverse when they are presented in a random order, and experiment 3: How humans can recognize random linear combinations of trained primitive patterns given by a set of discrete vibro-actuators on the forearm.

A Pico Vibe 10-mm vibration motor – 3 mm type (Precision Microdrives) in Figure 15.1A was used to make a wearable haptic-based pattern feedback system as shown in Figure 15.1B. There are seven Pico Vibe 10-mm vibro-actuators (model 310-103) arranged in equal distance (7 cm) in the array as shown in Figure 15.1B. The seven Pico Vibe 10-mm vibro-actuators are attached to the seven belts which can be adjusted by strapping securely to the arm of the different subjects. The different intensities for the vibrators are generated by an Arduino Mega motherboard and the amplitude was modulated by a simple power amplifier circuit as shown in Figure 15.1C. We recalibrated the vibro-actuators at different voltage inputs by actuating them on an ATI 6-axis Mini force sensor. We noticed that the linear regression between the measured frequency of vibration and input voltage had a coefficient of determination (R^2) of 0.94. The amplitude of vibration also had a positive trend against the input voltage, but at $R^2 = 0.34$. Therefore, in the rest of the paper, we can reliably state that subjects felt a proportional change of the frequency of vibration accompanied by a weak change of amplitude of vibration. Moreover, frequency control in vibro-actuators helps to keep the attention of the subject during persistent stimulation as opposed to using an alternative like magnitude or pressure.

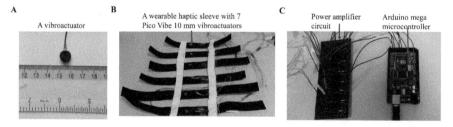

Figure 15.1 Hardware design for wearable haptic sleeve: (A) Pico Vibe 10-mm vibro-actuator (model 310-103), (B) Wearable vibro-tactile actuator arrays with seven Pico Vibe 10-mm vibro-actuator motors, and (C) Arduino Mega motherboard and power amplifier circuit to generate different intensity patterns.

The experimental protocol was approved by the King's College London Biomedical Sciences, Medicine, Dentistry, and Natural & Mathematical Sciences research ethics committee.

15.2.1 Haptic Primitive Templates Generation

The Gaussian templates were generated by the standard MATLAB function (MATLAB 2014b),

$$y = gaussmf(x, [sig, c]) \tag{15.1}$$

where $sig = std$, and c is the center of the distribution. The sig for pattern T, TR, TL, and THA are 1 and 0.5 for THS, respectively. Moreover, the amplitude of the THA was maintained at half of the rest of the templates.

15.2.2 Experimental Procedure

Subjects wore the haptic-based pattern feedback sleeve as shown in Figure 15.2A. Subjects were asked to keep the arm outstretched during the

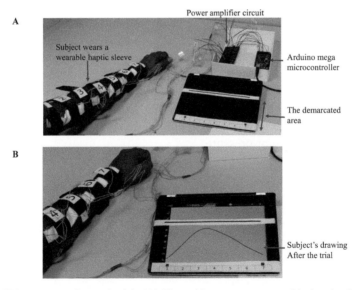

Figure 15.2   An experimental trial: (A) The subjects wear a wearable haptic sleeve with seven Pico Vibe 10-mm vibration motors. The drawing area is demarcated and used hardware is shown on Figure 15.2A, and (B) The subject drew the intensity felt from the vibro-tactile actuator arrays was during the trial on ipad. Draw free app (Apple Inc.) software is used as a drawing tool.

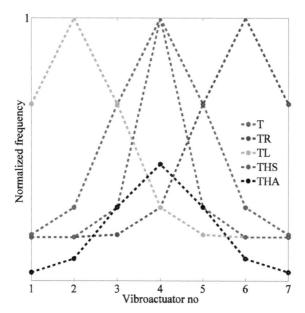

Figure 15.3 The templates for experiment 1 and experiment 2: T-Gaussian template. The patterns was generated by gaussmf (MATLAB R2012b) with a standard deviation 1 for T, TR, TL, and THA. The standard deviation 0.5 was used for THS.

experiments. The smooth curves of Figure 15.3 were selected as templates to generate different stimulation patterns. During a trial, all vibro-actuators vibrate simultaneously. Single trial ran for an average of 80 ms. During the first five trials, the template (T) in Figure 15.3 was played. Before playing each template, subjects were shown the printed template. Subjects were asked to draw a smooth curve representing what they perceive on an ipad sketching app [Draw free app (Apple Inc.)] after each trial as shown in Figure 15.2B. A drawing area on the ipad was clearly demarcated to match the size of the printed template as shown in Figure 15.2B. Just after the drawing, the next stimulation was given. On average, the inter-stimulus interval was 82 ms on average with 8.3 ms standard deviation. On average, the experiment stimulation was limited to 80 ms. After that period, subjects were free to start the drawing the estimated template.

15.2.3 Data Processing and Statistical Analysis

The same available pencil in the Draw free app was used for drawing throughout the experiments. Get Data Graph Digitizer version 2.6 was used to digitize the data (16 digits) on drawn lines. To obtain the regression coefficients, the respective template was generated by MATLAB 2014a with the exact length of the drawing curve for each trial. The regression coefficients were calculated between humans' sketch data (raw data) and the respective templates.

15.2.4 Experiment 1: To Understand How Humans Generalize a Gaussian Pattern in Scaling and Shifting

Ten healthy naive subjects (six male and two female) age between 24–39 participated in experiment 1. Experiment 1 was conducted to test how humans generalize a primitive template pattern (T) with respect to scaling and shifting. Shifting was done by left shift (TL) and right shift (TR), not up or down and scaling was done by shrinking (THS), and half in magnitude (THA) as shown in Figure 15.3.

In experiment 1, subjects were asked to wear the vibro-actuator belt. Subjects were trained and shown only printed template T. They were informed that the intensities of the vibro-actuators in the belt are directly proportional to the height of the template T during the stimulation. Subjects were told that they are supposed to draw a smooth curve with heights directly proportional to the intensities of the vibro-actuators after the stimulation. During the experiment, subjects were trained for only template T. Therefore, at the beginning of the experiment, template T was played five times. Then subjects were informed that trained and untrained templates would be played pseudorandomly during the experiment. Therefore, after first five training trials, TL, TR, THS, and THA patterns were played randomly. However, there are training blocks in between other templates to train template T as shown in Table 15.1. During the training blocks, subjects were informed that the printed template was shown prior to the trial. For all training sessions (when T played), the visual cue was provided. At the end of each trial, subjects were asked to draw what they felt during the trial. Pattern T repeated four times like in a block as shown in Table 15.1. Likewise, four blocks of templates were played during the experiments after first five trials. The rest of four intensity patterns (TR, TL, THA, and THS) played six times each during the experiment randomly as shown in Table 15.1. Therefore, subjects participated in 45 trials during the experiments. For more clarity, the trial number and respective played patterns are shown in Table 15.1.

Table 15.1 Intensity (Hz) for different templates and the order for experiment 1 and experiment 2(a). Rows represent each block of trials in experiments

	Vib. 1	Vib. 2	Vib. 3	Vib. 4	Vib. 5	Vib. 6	Vib. 7	Exp 1: Order	Exp 2: Order
T	103.33	140.60	281.96	400.00	281.96	140.60	103.33	1–5, 11–14, 20–23, 29–32, 38–41	1–5, 21, 29, 32, 34, 38, 41
TR	100.00	100.10	103.33	140.60	281.96	400.00	281.96	6, 10, 18, 27, 33, 45	6–10, 23, 27, 31, 33, 39, 42
TL	281.96	400.00	281.96	140.60	103.33	100.10	100.00	7, 17, 24, 35, 37, 44	16–20, 22, 26, 28, 35, 37, 44
THS	100.00	100.10	140.60	400.00	140.60	100.10	100.00	9, 16, 26, 28, 34, 42	THS was dropped in Ex 2
THA	51.66	70.30	140.98	200	140.98	70.30	51.66	8, 15, 19, 25, 36, 43	11–15, 24, 25, 30, 36, 40, 43

15.2.5 Experiment 2(a): To Understand How Humans can Recognize Trained Templates When they are Presented in a Random Order

The second experiment was conducted to test how subjects recognize all trained haptic feedback patterns when they are presented in a random order. Eight healthy naive subjects (six male and two female) aged 24–28 participated in the experiment 2(a). All instructions in experiment 1 were given to the subjects. However, not only the template T but also the templates TR, TL, and THA were shown to the subjects. Since the subjects were not able to distinguish the pattern THS from other patterns in experiment 1, the pattern THS was dropped and only patterns TL, TR, THA, and T were considered for experiment 2(a). During the experiment, first 20 trials were designed to train the subjects to learn the patterns T, TR, TL, and THA. Each training pattern was played five times. During those 20 training trials, subjects were shown the pattern to be played. Finally, the subjects were asked to draw a smooth curve representing the vibro-actuator intensity pattern they felt on an ipad screen. During the testing session, the four training templates were played in pseudorandom order to achieve counter balancing. Each template was played six times making the total number of trials experienced by each subject to be 44. For more clarity, the order of the trials and the frequencies of the vibro-actuators are shown in Table 15.1.

15.2.6 Experiment 2(b): To Understand How Humans Can Recognize Trained Inverse Templates When They are Presented in a Random Order

Experiment 2(b) was designed to test how subjects recognize inverse of the trained templates in experiment 2(a). Eight healthy naive same group of subjects in experiment 2(a) (six male and two female) aged 24–28 participated in the experiment 2(b). Subjects were informed that the inverse of the templates in experiment 2(a) is used during the experiment 2(b). However, for simplicity, only inverse templates of T, TL, and TR were used. Here after we use IT, ITL, and ITR, respectively, for denoting the inverse templates of T, TL, and TR. Again first 15 trials were designed to train the subjects to learn the templates of IT, ITL, and ITR. Each training pattern was played five times. The same procedure was followed as experiment 2(a). After the training session, trained three inverse patterns were played four times randomly as shown in Table 15.2. Therefore, subjects participated in 27 trials. For more

Table 15.2 Experiment 2(b): Trial order for testing inverse templates of experiment 2(a)

Template	Trial Order
IT	1–5, 17, 20, 24, 25
ITL	6–10, 16, 18, 21, 23
ITR	11–15, 19, 22, 26, 27

clarity, the order of the trials is shown in Table 15.2. However, counter balanced was followed as experiment 2(a).

15.2.7 Experiment 3: To Understand How Humans can Recognize Random Linear Combinations of Trained Primitive Templates Given by a Set of Discrete Vibro-Actuators on the Forearm

Experiment 3 was designed to understand how humans recognize linear random combinations of trained primitive patterns given by a set of vibro-actuators on the forearm. Subjects were informed that untrained new templates are played during the stimulation. The same group of subjects in experiment 2 participated in experiment 3 just after experiment 2. Therefore, no training sessions were conducted for experiment 3. The patterns were

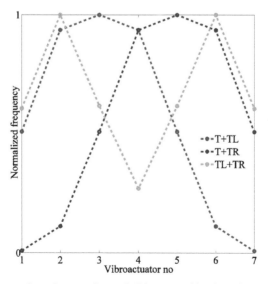

Figure 15.4 The templates for experiment 3: Linear combination of two primitive patterns: The combination of primitive patterns of T + TR, T + TL, and TL + TR is shown.

Table 15.3   Experiment 3: Intensity (Hz) and pattern order for combined templates

Intensities (Hz):	Vib. 1	Vib. 2	Vib. 3	Vib. 4	Vib. 5	Vib. 6	Vib. 7	Trial Order
T+TL	203.66	374.37	400.00	374.37	203.66	44.73	3.66	1, 5, 7, 12, 15
T+TR	3.66	44.73	203.66	374.37	400.00	374.37	203.66	3, 6, 8, 10, 14
TL+TR	242.53	400.00	246.97	108.23	246.97	400.00	242.53	2, 4, 9, 11, 13

selected as linear combinations of T + TL, T + TR, and TL + TR as shown in Figure 15.4. Moreover, the frequencies of the combined patterns and order of the played patterns are shown in Table 15.3. It was found that a perceptible range by humans is 20–400 Hz [16]. Therefore, frequencies were normalized to bring the maximum frequency to 400 Hz as shown in the values in Table 15.3 to attain maximum sensitivity.

15.3 Results

15.3.1 Experiment 1

The sketched raw data for the pattern T, TL, TR, THS, and THA in experiment 1 are shown in Figure 15.5. The template patterns are shown by a black dashed line. The sketched data in Figure 15.5 were regressed by respective templates shown by a black dashed line for each template. The average regression coefficients are shown in Figure 15.6. The results in Figure 15.6A show high correlationship between the data and template only when the template T was played. For clarity, the variation of regression coefficients in each training blocks of template T in Table 15.1 is shown in Figure 15.6B. For more clarity, the fitted curves for each block are shown on top of the respective curves. The results show that only block 1 has a negative trend while blocks 2–5 have positive trends in learning during the trials. The subsequent training blocks (2–5), in Figure 15.6B show a positive trend indicating that there is a relearning effect after being exposed to patterns other than T. The regression coefficients of the first five trials in Figure 15.6A confirm that this training block sets up a baseline. The negative trend from blocks 2–5 would come from interference from other randomly played templates during the testing trials.

Moving to the regression coefficients in templates TL, TR, THA, and THS, regression coefficient values are relatively low when the sketched data are regressed with respective templates as shown in Figure 15.6C. Moreover, it is noticed that some regression values are less than 0 for templates TR and TL in Figure 15.6C. The deviation can be noticed in humans' sketched

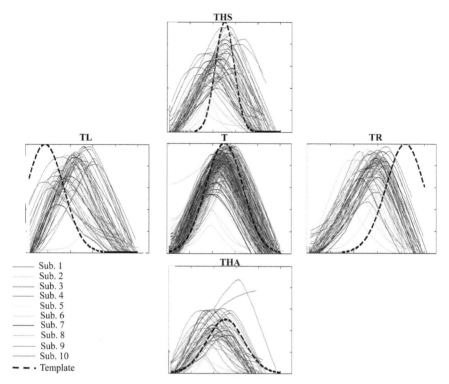

Figure 15.5 The raw data for experiment 1: The sketched data for pattern T, TL, TR, THS, and THA for all trials. The all templates are shown by a black dashed line.

data in Figure 15.5 for templates TR and TL. However, the low and negative regression coefficients and higher variability values in Figure 15.6C suggest that subjects were not able to shift the pattern (TL and TR) they are trained in. This might come from the fact that the memory of the pattern T interferes with subjects' perception as shown in raw data in Figure 15.5. For more clarity, the data were regressed with template T as shown in Figure 15.6D. (Note that in Figure 15.6C the regression was done against the actual pattern that was played).

Figure 15.6D shows the improvement of the regression coefficients for the patterns when the data are regressed with template T, with respect to Figure 15.6C for pattern TL and TR. The improvement of regression coefficient might come from the interference of trained pattern T. Moreover, none of the regression coefficients are < 0 in Figure 15.6D as noticed in Figure 15.6C. This again suggests interference of the trained pattern T in subject's memory

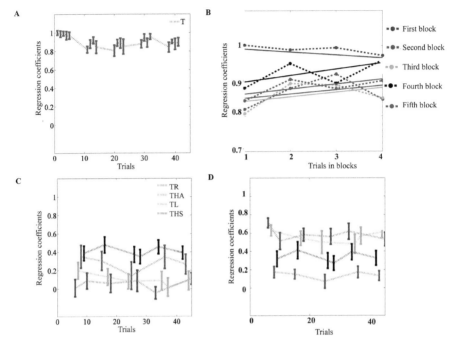

Figure 15.6 Experiment 1: Average regression coefficients (in this experiment, subjects were trained only for template T): (A) Regression coefficients for training template T over the trials, (B) Average regression coefficient values for training blocks 1–5 in Table 15.1. The fitted curves for each block are shown on top of the respective curve to understand the positive (only block 1) and negative (from block 2–5) trends in learning during the trial, (C) Testing session regression coefficients for templates TL, TR, THS, and THA when the data are regressed with its respective templates. (D) Regression coefficients for templates TL, TR, THS, and THA when the data are regressed with template T.

in shifting as noticed in Figure 15.6C. For more clarity, Mann-Whitney U test ($\alpha = 0.05$) was conducted to test whether the regression coefficients in Figures 15.6C and 15.6D are statistically significantly different in all patterns (Since data were not normally distributed, non-parametric Mann-Whitney U test was used to test the significance). It is shown that regression coefficients for pattern TR ($p = 0.002$) and TL ($p = 0.002$) are significantly different. However, for THA, $p = 0.055$, and for THS, $p = 0.132$ suggesting that they are not significantly different at 95% confidence level. The significance test results in TL, and TR again provide an evidence of interference of the trained pattern T in subject's memory in shifting T to left or right. The regression coefficients in Figures 15.6C and 15.6D for THS and THA suggest that THS and THA have different structures whereas the former two are of the same structure of T other than the shift. However, experiment 2 was conducted to train not only the template T but also all templates. Moreover, the inverse templates of the experiment 1 were also trained. The higher significance test value and regression coefficients of THS suggest that interference mostly involved shifting than scaling. Moreover, it might suggest that THS is statistically independent. Therefore, for simplicity the pattern THS was dropped in experiment 2.

15.3.2 Experiment 2(a)

The sketched raw data for the pattern T, TL, TR, and THA in experiment 2(a) are shown in Figure 15.7. The templates are shown by black dashed line. The sketched data in Figure 15.7 were regressed against respective templates. The average regression coefficients are shown in Figure 15.8 when subjects were trained and tested in a random order of patterns T, THA, TL, and TR in experiment 2(a). In general, in Figure 15.8A, all regression coefficients have improved with respect to Figure 15.6C for all templates. The average

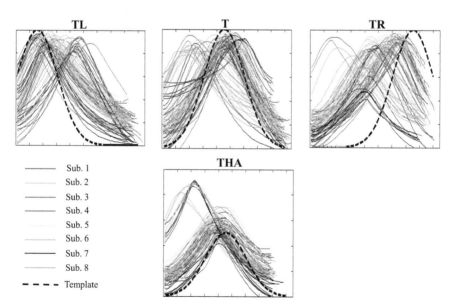

Figure 15.7 The raw data for experiment 2(a): The sketched data for pattern T, TL, TR, and THA for all trials in experiment 2(a). The templates are shown by a black dashed line. The template THS in experiment 1 was dropped in experiment 2(a).

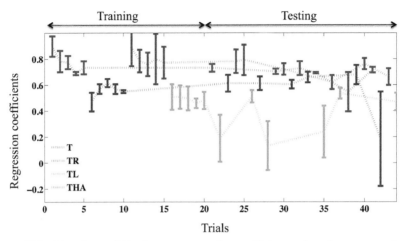

Figure 15.8 Experiment 2(a): Average regression coefficients when data regressed with respective template in Figure 15.3. Standard deviations of the regression coefficients are shown by error bars.

regression coefficients of training session are higher for T and THA with respect to TL and TR as shown in Figure 15.8A. It implies that subjects have a better ability to recognize scaled template than shifted ones as noticed in experiment 1. Moreover, it can be seen in sketched data in Figure 15.7 too. Interestingly, subjects can recognize these four templates when they are played in random order, provided they were trained earlier. Therefore, those results show that subjects can recognize trained primitive patterns when the vibro-actuator array generates different stimulations. The experiment 2(b) was conducted to understand how subjects recognize inverse templates when they are trained.

15.3.3 Experiment 2(b)

Average regression coefficients when subjects are trained for inverse templates of T, TL, and TR in experiment 2(a) are shown in Figure 15.9. The regression coefficients for training and testing sessions are shown in Figure 15.9 when they were regressed against by respective templates. The results show that there is no any improvement in regression coefficients after training the subjects as we noticed in experiment 1 and experiment 2(a). To visualize actual data, the raw data were plotted as shown in Figure 15.10. For more clarity, the corresponding experimental trials for its regression

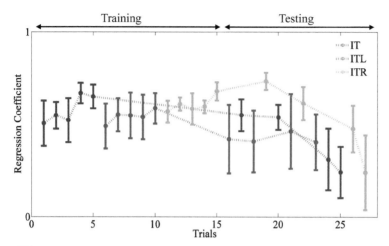

Figure 15.9 Experiment 2(b): Average regression coefficients when the sketched patterns were regressed against templates IT, ITL, and ITR. Training and testing sessions are shown in the figure.

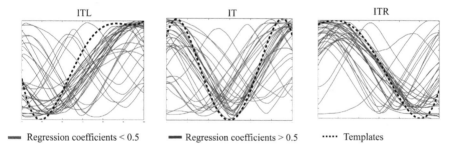

Figure 15.10 The raw data for experiment 2(b) for templates IT, ITL, and ITR. Corresponding regression coefficients <0.5 and >0.5 are shown by a red and blue colors respectively. Templates are shown by a black dashed line.

coefficients <0.5 and >0.5 are shown by red and blue, respectively, in Figure 15.10. Moreover, the templates are shown by a black dashed line. Figure 15.10 shows that most of the regression coefficients values are <0.5 as shown by red confirming subjects were not able to recognize inverse patterns as we noticed low regression coefficients in Figure 15.9. This might come from subjects find it difficult to learn inverse patterns because they are bimodal patterns [e.g., the shape of IT. (please refer template IT in Figure 15.10 by black dashed line)].

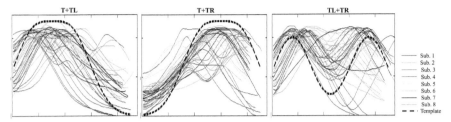

Figure 15.11   Experiment 3: The human sketched data for pattern T+TL, T+TR, and TL+TR for all trials. The templates are shown by a black dashed line.

15.3.4 Experiment 3

The sketched raw data for the pattern T+TL, T+TR, and TL+TR in experiment 3 are shown in Figure 15.11. The templates are shown by a black dashed line. The sketched data in Figure 15.11 were regressed by respective templates. The results of experiment 3 are shown in Figure 15.12 when the sketched data were regressed with the respective templates in Figure 15.4. The average regression coefficient values are higher for T+TL and T+TR as shown in Figure 15.12. This might come from the frequency values of T+TL and T+TR patterns as shown in Table 15.3. In Table 15.3, five adjacent vibro-actuators are in humans' threshold frequency range. Subjects were not able

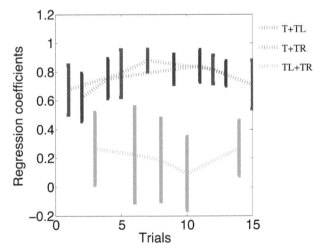

Figure 15.12   Experiment 3: Regression coefficients when data regressed with respective template in Figure 15.4. Standard deviation of the regression coefficients are shown by a error bars.

to recognize when not enough vibro-actuators fall within the most sensitive frequency range [16] as noticed in TL+TR. The low regression coefficients and higher variability values suggest that activating sufficient adjacent vibro-actuators together within the threshold frequency range is more appropriate to convey spatial information. Moreover, subjects might be able to recognize uni-model than bi-model as shown in Figure 15.11. These results in experiment 3 suggest that the vibro-actuator array has the potential to convey spatial information using uni-modal patterns to humans. The results in experiment 2(b) show that the efficiency of this mode of communication can be improved by having a higher number of adjacent vibro-actuators as we observed in results in experiment 3.

Significance test was carried out to see if each template was the best predictor of the response compared to the other templates in the same set for experiment 2(a) and experiment 3. This would provide an indication for how distinguishable the patterns are. The results can be summarized as follows for experiment 2: $P_{(T,L)} < 0.05$, $P_{(T,TR)} < 0.05$, $P_{(THA,TL)} < 0.05$, $P_{(TR,THA)} < 0.05$, and $P_{(THA,T)} > 0.05$. The significance results in experiment 2 suggest that except the pair (T, THA), all other pairs are significantly different. Moreover, significance test results in experiments 3 show that $P_{(TL+TR,T+TL)} > 0.05$ and $P_{(T+TL,T+TR)} > 0.05$.

15.4 Discussion

This chapter presents experimental evidence of the capabilities and limitations of the human somatosensory system to distinguish and recognize a class of primitive haptic feedback patterns presented after prior training. Three consecutive experiments that provide insights into how humans recognize trained cutaneous feedback patterns as well as their shifts and combinations have been discussed. The results in this paper provide new insights into an important area of tactile input to inform shape recognition. The subjects' drawings of stimulus waveforms were captured and regression coefficients were used to understand humans' ability to recognize the given stimulations by different templates via a vibro-actuator array.

A deeper understanding of humans' ability to generalize and recognize trained templates is important to convey some messages via vibro-actuator stimuli when humans have to work in noisy environments. Moreover, those trained templates could be considered as primitives of a haptic-based language. Those primitives and subjects' responses would give an idea as to how they can be used to increase humans' perception when they are in noisy environments like factory, search, and rescue. The results of experiments 1–3 provide us the kind of primitive shapes that could be used to train humans to use a template-based haptic-based communication language.

Results of experiment 1 show that even though subjects were able to scale, they find it difficult shift a trained template T. This suggests that the trained template T interferes with a shifted pattern. The raw data in experiment 1 show higher deviation from the played template. Even for the template T in Figure 15.5 has a higher deviation from the trained template T. This variability could come from physiological factors like muscle tension and psychological factors like attention.

These evidence of interference raise further questions about how haptic sensory memory can cause after effects as noticed in auditory sensory memory in previous studies [26, 27]. Therefore, future studies can investigate the effect of longer intertril breaks and even the effect of an overnight break on the ability of subjects to recognize a learned template as well as the interference of a learned template on a different template. Studies on motor memory consolidation suggest that the robustness of motor memories increases for breaks longer than 5 min and that clear memory consolidation effects can be seen after a 4-h break [28].

However, results in experiment 2(a) show that average regression coefficients have improved with respect to experiment 1. These improvements show that subjects have an ability to recognize the same distributed haptic pattern (in this case T) even when it is shifted along the arm. Here TL is more precise and less strongly shifted to the center of the TR. It could come from more sensitivity bias in shoulder than elbow (please note vibro-actuators labeled from 1 to 7 in Figure 15.3 are from the upper arm to the lower arm) as shown in raw data in Figure 15.7. It would be useful to further investigate the distribution of sensitivity to vibro-tactile feedback along the human arm. Even though the same procedure was used for training for all experiments, the low regression coefficients in Figures 15.6 and 15.8 suggest that even if recognition of the tactile patterns were good, performance would still be poor if drawing the visual representation was difficult. The degree of drawing difficulty may be quantified in the future by asking the subjects to draw a pattern visually presented to them immediately before sketching. The regression values between the sketch and the template can be used to scale the regression values obtained for sketched patterns and their corresponding templates given on the skin.

From the results of experiment 2(b), together with the results for a linear combination of two uni-modal patterns giving rise to a bi-modal pattern, we can conclude that subjects find it difficult to recognize bi-modal patterns even if they are trained a priori. This may arise from a somatosensory mechanism to interpolate bimodal patterns to uni-modal patterns even if a uni-modal pattern was not trained before hand.

The combination of templates was simply a sum of two templates presented using the seven vibro-actuators in experiment 3. In terms of location, the two constituent patterns that went into the combined pattern were not shifted. Therefore, if the somatosensory memories of the two constituent patterns can be retrieved and combined, the participants should be able to reconstruct the combined pattern though it was not experienced before. Instead, we see that the two humps in the combined pattern were not accurately reconstructed. Our interpretation is that the two constituent templates fell within a minimum distance needed to distinguish two distinct humps. This provides useful design guidelines for a future attempt to use primitive templates to construct complex messages.

Higher regression coefficients for combined primitive templates in experiment 3 show that subjects could recognize linear combinations of patterns T+TL and T+TR. In T+TL and T+TR, provided five adjacent vibro-actuators are in humans' threshold frequency range (humans' most sensitive frequency range is 200–400 Hz [16]). Subjects were not able to recognize linear combinations of trained templates when not enough vibro-actuators fell within the most sensitive frequency range [16] as noticed in TL+TR. These results also show that bi-modal patterns resulting from linear combinations of trained uni-modal patterns can interfere with trained uni-modal patterns falling between the two peaks of the bi-modal pattern as we noticed in experiment 2(b). This too can be explained by the locality of somatosensory representation seen in the above results.

In summary, our results show that (1) subjects find it easier to recognize unimodal haptic templates, (2) subjects tend to generalize bi-modal vibro-tactile patterns to uni-modal patterns at least in the range of separation of the two peaks in these experiments, and (3) haptic memory is location specific in the sense that a template trained on a given area of the arm is best recognized when played at the same location. When the same template is shifted along the arm, the original template interferes with the shifted pattern. This allows us to train different patterns in different areas of the arm; 4) subjects can recognize linear combinations of previously trained haptic templates. This provides the opportunity to build a haptic-based communication language that can encode complex messages by linearly combining primitive templates like "soft" + "obstacle." The results explain as to how to use cutaneous feedback to the humans to convey some messages when humans working in noisy environments. It would help them to mentally construct the message by their training experiences as noticed in the results in experiments. In this paper, we show how scaling/shifting of a Gaussian template is recognized. The findings would give an insight about how special haptic memory is represented in the brain, how they could be linearly combined, and how humans can be trained for using multiple haptic patterns to decode complex messages.

In the future, it would be interesting to test whether subjects could decode the messages from the wearable haptic sleeve when the user is mobile and active. It will inform to develop a vocabulary to be used in a haptic language for special information and to design training protocols. In this regard, studies in [29] show how different age groups differentiate tactile perceptions. Therefore, future studies could try to identify differences across different age groups, prior experience of doing similar tasks, and even gender in learning a haptic template based language.

Acknowledgment

The authors would like to thank UK Engineering and Physical Sciences Research Council (EPSRC) grant no. EP/I028765/1, EP/NO3211X/1, and EP/N03211X/2, the Guy's and St Thomas' Charity grant on developing clinician-scientific interfaces in robotic assisted surgery: translating technical innovation into improved clinical care (grant no. R090705), and Vattikuti foundation.

References

[1] Hale, K. S., and Stanney, K. M. (2004). "Deriving haptic design guidelines from human physiological, psychophysical, and neurological foundations," in *Proceedings of the IEEE Computer Graphics and Applications,* 24, 33–39.

[2] Thoroughman, K. A., and Shadmehr, R. (2000). Learning of action through adaptive combination of motor primitives. *Nature* 407, 742–747.

[3] Gilson, R. D., Redden, E. S., and Elliott, L. R. (2007). *Remote Tactile Displays for Future Soldiers (No. ARL-SR-0152).* Orlando: University of Central Florida.

[4] Jones, L. A., and Lederman, S. J., (2006). *Human Hand Function.* Oxford: Oxford University Press.

[5] Tsukada, K., and Yasumura, M. (2004). "Activebelt: Belt-type wearable tactile display for directional navigation," in *Proceedings of the UbiComp 2004: Ubiquitous Computing* (Berlin: Springer), 384–399.

[6] Scheggi, S., Aggravi, M., Morbidi, F., and Prattichizzo, D. (2014). "Cooperative human-robot haptic navigation," in *Proceedings of the IEEE International Conference on In Robotics and Automation (ICRA)*, 2693–2698.

[7] Pielot, M., Poppinga, B., Heuten, W., and Boll, S. (2011). "A tactile compass for eyes-free pedestrian navigation," in *Proceedings of the Human-Computer Interaction–INTERACT* (Berlin: Springer), 640–656.

[8] Pielot, M., Poppinga, B., and Boll, S. (2010). "PocketNavigator: vibrotactile waypoint navigation for everyday mobile devices," in *Proceedings of the 12th International Conference on Human Computer Interaction with Mobile Devices and Services*, 423–426.

[9] Velazquez, R., Fontaine, E., and Pissaloux, E. (2006). "Coding the environment in tactile maps for real-time guidance of the visually impaired," in *Proceedings of the IEEE International Symposium on In Micro-NanoMechatronics and Human Science*, 1–6.

[10] Dakopoulos, D., and Bourbakis, N. (2009). "Towards a 2D tactile vocabulary for navigation of blind and visually impaired," in *Proceedings of the IEEE International Conference on In Systems, Man and Cybernetics SMC*, 45–51.

[11] Bliss, J. C., Katcher, M. H., Rogers, C. H., and Shepard, R. P. (1970). "Optical-to-tactile image conversion for the blind," in *Proceedings of the IEEE Transactions on Man-Machine Systems*, 11(1), 58–65.

[12] Wall III, C., Weinberg, M. S., Schmidt, P. B., and Krebs, D. E. (2001). "Balance prosthesis based on micromechanical sensors using vibrotactile feedback of tilt," in *Proceedings of the IEEE Transactions on Biomedical Engineering*, 48, 1153–1161.

[13] Priplata, A. A., Niemi, J. B., Harry, J. D., Lipsitz, L. A., and Collins, J. J. (2003). Vibrating insoles and balance control in elderly people. *The Lancet* 362, 1123–1124.

[14] Rupert, A. H. (2000) An instrumentation solution for reducing spatial disorientation mishaps. *IEEE Eng. Med. Biol. Mag.* 19, 71–80.

[15] Oakley, I., Kim, Y., Lee, J., and Ryu, J. (2006). "Determining the feasibility of forearm mounted vibrotactile displays," in *Proceedings of the IEEE In Haptic Interfaces for Virtual Environment and Teleoperator Systems, 14th Symposium*, 27–34.

[16] Gemperle, F., Hirsch, T., Goode, A., Pearce, J., Siewiorek, D., and Smailigic, A. (2003). *Wearable Vibro-tactile Display*. Technical report, Carnegie Mellon University.

[17] Summers, I. R., Whybrow, J. J., Gratton, D. A., Milnes, P., Brown, B. H., and Stevens, J. C. (2005). Tactile information transfer: A comparison of two stimulation sites. *J. Acoust. Soc. Am.* 118, 2527–2534.

[18] Kaaresoja, T., and Linjama, J. (2005). "Perception of short tactile pulses generated by a vibration motor in a mobile phone," in *Proceedings of the Eurohaptics Conference and Symposium on Haptic Interfaces for Virtual Environment and Teleoperator Systems, World Haptics.*

[19] Van Erp, J. B. (2002). "Guidelines for the use of vibro-tactile displays in human computer interaction," in *Proceedings of the Eurohaptics*, 18–22.

[20] Stepanenko, Y., and Sankar, T. S. (1986). Vibro-impact analysis of control systems with mechanical clearance and its application to robotic actuators. *J. Dynamic Syst. Measurement Contr.* 108, 9–16.

[21] Benali-Khoudja, M., Hafez, M., Alexandre, J. M., Khedda, A., and Moreau, V. (2004). "VITAL: a new low-cost vibro-tactile display system," in *Proceedings of the IEEE International Conference in Robotics and Automation*, 1, 721–726.

[22] Zaitsev, V., and Sas, P, (2000). Nonlinear response of a weakly damaged metal sample: a dissipative modulation mechanism of vibro-acoustic interaction. *J. Vibr. Contr.* 6, 803–822.

[23] Piateski, E., and Jones, L. (2005). "Vibrotactile pattern recognition on the arm and torso," in *Proceedings of the IEEE In Eurohaptics Conference and Symposium on Haptic Interfaces for Virtual Environment and Teleoperator Systems, World Haptics* 2005, 90–95.

[24] Shimizu, Y., Saida, S., and Shimura, H. (1993). Tactile pattern recognition by graphic display: Importance of 3-D information for haptic perception of familiar objects. *J. Percept. Psychophys.* 53, 43–48.

[25] Behrmann, M., and Ewell, C. (2003). Expertise in tactile pattern recognition. *J. Psyc. Sci.* 14, 480–492.

[26] Näätänen, R., Paavilainen, P., Alho, K., Reinikainen, K., and Sams, M. (1989). Do event-related potentials reveal the mechanism of the auditory sensory memory in the human brain? *Neurosci. Lett.* 98, 217–221.

[27] Sams, M., Hari, R., Rif, J., and Knuutila, J. (1993). The human auditory sensory memory trace persists about 10 sec: neuromagnetic evidence. *J. Cogn. Neurosci.* 5, 363–370.

[28] Shadmehr, R., and Holcomb, H. H. (1997). Neural correlates of motor memory consolidation. *Science* 277, 821–825.

[29] Reuter, E. M., Voelcker-Rehage, C. S., Vieluf, A. H. W., and Godde, B. (2013). A parietal-to-frontal shift in the P300 is associated with compensation of tactile discrimination deficits in late middle-aged adults. *Psychophysiology* 50, 583–593.

16

RobinHand Haptic Device

Krzysztof Lis, Łukasz Mucha, Krzysztof Lehrich and Zbigniew Nawrat

Prof. Z. Religa Foundation of Cardiac Surgery Development, Zabrze, Poland

Abstract

This chapter presents the stages of the design process and the working principle of the RobinHand control device adapted for the needs of the STIFF-FLOP project. The chapter also presents the concept and method of transferring tactile sensations (force and vibro-tactile feedback) from a real device or virtual reality to the user. The authors also discuss the structure, the working principle, and the application of the interface. A number of prototypes are developed and presented along with a brief description of their structure.

16.1 Introduction

There is a great potential in using medical robots in surgery as they offer increased precision and enable minimally invasive access to the operating field. Medical surgery typically makes use of tele-manipulators, with a human who takes decisions concerning the motion and tasks to be performed on the one side of the robot and a surgical instrument executing tasks in the working space inside a patient's body on the other side. Control of the medical robots takes place in real time. The trajectories are defined typically in the Cartesian space or in the configuration space variables (coordinates). The main problems involved in controlling a tracking motion include ensuring the required precision and stability of motion.

While designing a control device and a control algorithm, it is necessary to take into account variable working conditions that result from the performance of different tasks [1–3]. The simplest control system may be applied when the motion controller's kinematics and the robot's kinematics are similar; in this situation, the motion of the motion controller directly corresponds to the robot's motion. Optimal kinematics of the manipulator (slave) typically differs from the kinematics of the motion controller (master). It requires calculations of the control system based on forward and inverse models of the tool robot and the motion controller. The system may be equipped with detection, processing/conversion, and transmission modules, which relay feedback information, reflecting the interaction of the tool with objects in the operating field to the operator in a number of ways including force feedback, optical, thermic, and vibratory sensations. Both signals carry the information about the actions of the operator [4–6]. In surgical robots, a lack of effective force feedback makes the surgeon's work significantly more difficult. The surgeon can rely only on visual feedback to safely interact with the body's tissues. In order to eliminate the deficiencies resulting from this type of control, haptic (Greek: háptein – attach, grab) motion controllers are developed, which allow controlling the robot while providing the operator with subjective sensations of direct contact with the manipulated object at the same time [5–9].

16.2 The User Interface RobinHand

For the purposes of the STIFF-FLOP project, three versions of the motion controller called *RobinHand* have been developed and adapted. The first of them was developed and made in 2015 by a group led by Krzysztof Lis. The *RobinHand H* motion controller has three degrees of freedom and does not offer any feedback. This motion controller is intended to control the visual tracking robot and manipulating the robot in a virtual 3D reality. *RobinHand H* interface is presented in Figure 16.1 [4, 5].

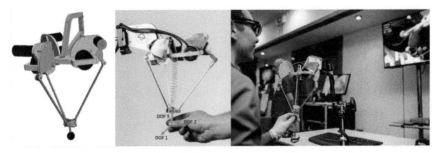

Figure 16.1   Haptic *RobinHand H*: CAD model, degrees of freedom, and system operation in a virtual 3D environment [5].

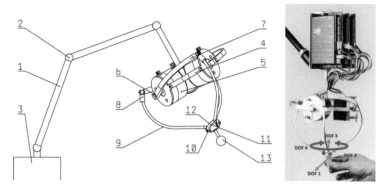

Figure 16.2   A perspective view of the manipulator *RobinHand F* [5, 10].

The second version – *RobinHand F* is equipped with drives which offer force feedback in the first three degrees of freedom (DOF 1–3) and the third version with a rotational degree of freedom (DOF 4) without force feedback (Figure 16.2).

In this project (Figure 16.2), the manipulator comprises a fixing arm (1) with any number of joints (2) enabling angular deflection and adjusting individual parts of the mounting system. The fixing arm (1) is mounted to the base (3) in any way on the one side, and to the fixed platform (4) on the other side. There are three motors with built-in encoders (4) mounted on the fixed platform (5). Lines comprising rotational axes of shafts (6) of the three motors (5) intersect at one point at an angle of 120°. There are first arched connectors (7) fixed on the shafts (6) of the motors (5) with an angle range of 90°. The first arched connectors (7) are rigidly connected with their first ends to the shafts (6) of the motors (5), thus enabling their rotation, which results in an angular deflection of the second end of the connectors (7). The second ends of the first arched connectors (7) are connected by means of joints (8) with three degrees of freedom with the first ends of the second arched connectors (9). The three degrees of freedom are a result of using a pivot-type joint (8) between the first arched connector (7) and the second arched connector (9). The angle range of 90° of the first arched connectors (7) requires such a shape of the connector that the rotational axis of the joint (8) in relation to the connector (7) is at an angle of 90° in relation to the rotational axis of the shaft (6) (a top view of the fixed platform with an indicated working angle is presented in Figure 16.3). The second ends of the second arched connectors (9) have transverse arched arms (Figure 16.4) and are connected with the arms by means of a pivot joint (10) to a mobile platform 11. On the mobile

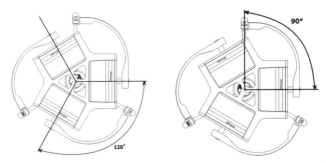

Figure 16.3   A top view of the fixed *RobinHand F* platform with an indicated work angle [10].

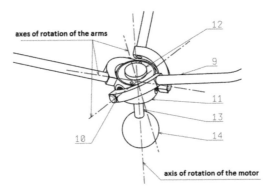

Figure 16.4   A perspective view of the mobile platform with indicated rotational axes *RobinHand F* [10].

platform (11), there is a motor mounted inside with a built-in encoder (12). A control knob (14) for the operator or alternatively other sub-assemblies of the controller allowing the operator to increase the number of degrees of freedom is mounted to the shaft (13) of the motor (12). The rotational axes of the second ends of the second arched connectors (9) intersect at one point on the line comprising the rotational axis of the shaft (13) of the auxiliary motor (12) mounted on the mobile platform (11). Figure 16.5 shows the kinematics of motion of the individual kinematic pairs.

The STIFF-FLOP project was evolving along with the growing experience of the FCSD team. In the next version of the motion controller, the design of the knob (held by the operator while using the controller) was modified. *RobinHand L* version was adjusted to the needs of the project, offering 7 DOF, where the first three degrees used force feedback (Figure 16.6).

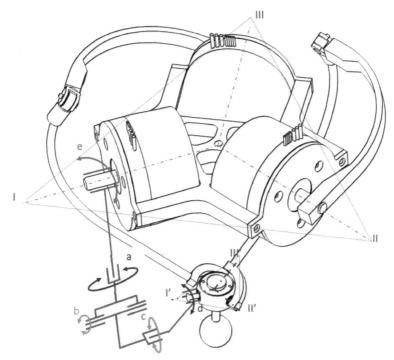

Figure 16.5 A perspective view of the manipulator with indicated kinematics of motion of individual kinematic pairs *RobinHand F* [10].

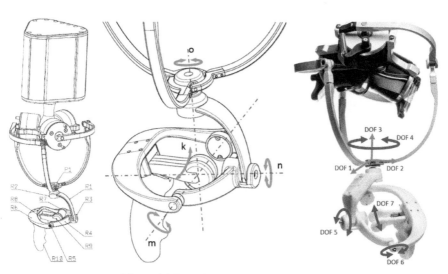

Figure 16.6 *RobinHand L* haptic platform.

The additional degrees of freedom allow controlling the tools/robot with additional articulated joints. The ergonomics of the manipulator was improved by adjusting it to the operator's hand so that the lines going through the articulated parts cross each other at the same point, just between the fingers of the operator. This solution made the handling of the controller more intuitive. To execute the construction of such a sophisticated form, it was necessary to implement the most advanced 3D printing technology as well as systems of joining different modules (metal-plastic and metal-composite). The use of rapid prototyping (FDM 3D printing with PC-ABS material) resulted in minimizing the weight of the component manipulated by the operator. This solution also contributed to reducing the time needed to develop other prototype versions to find the best match with the operator's needs [11]. Figure 16.6 shows the components made with R1, R4, R6, R9, and R10 methods. The electronics behind the control system were improved. This motion controller was made in two versions: mobile – fixed on the articulated arm (Figure 16.7a) and in a version which allows installation on the brand-new version of the STIFF-FLOP control console (Figure 16.7b). The above solutions are legally protected in the form of patent applications: US 9393688, EP 2990005, PL W.124541 [10, 12, 13]. The controller is currently used to

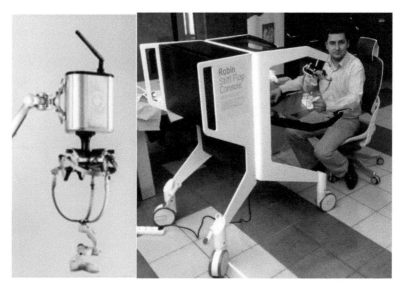

Figure 16.7   Haptic *RobinHand L* – mobile version and the version adapter for the STIFF-FLOP console.

control the *Robin Heart "Pelikan"* robot in Foundation Cardiac Surgery Development (FCSD) in Zabrze [14].

16.3 RobinHand in STIFF-FLOP Project

The implementation of the user interface RobinHand has been performed exploiting requirements and specifications based on a previously conducted Robin Heart project. Most importantly, we incorporated digital mapping allowing free movement of the surgeon's hand and providing haptic feedback to the surgeon's palm, relaying back the interaction of the robot arm's working tip with the inside of the abdominal cavity of the patient.

The movement performed by the operator's hand is captured by encoders in the haptic RobinHand unit. Information from the encoder is processed by a microcontroller (STM32) and is sent in the form of Cartesian coordinates to the microcontroller which operates the pressure valves used to control the robot arm's movements. Feedback information from the STIFF-FLOP is collected from the pressure sensors connected to each channel and is used to calculate the position error and in turn to bring (in order to bring) the arm to the desired position in the X, Y, Z space (Figure 16.8).

In order to check the haptic *RobinHand* functionality, FCSD tested and evaluated the integrated haptic console system (Figures 16.9 to 16.12). The system was launched and tested with force feedback acting both on the haptic console and the vibrating sleeve described in Chapter 14 [3, 15].

The integrated system used for the testing comprises the following:

- Two pneumatically operated robot arms equipped with both lateral flexi force sensor as well as frontal and circular lateral pressure sensor;
- A completely integrated system composed of the FCSD *RobinHand* haptic, a haptic vibration sleeve, and a soft robot arm inspired by STIFF-FLOP (made from Dacron –polyester fiber vessel prosthesis and Ecoflex – 30 silicone).

16.4 Operator–Robot Cooperation through Teleoperation and Haptic Feedback

In all the remote tests, a user (at the master site) provided input signals through a user-input console. The provided inputs were then transmitted via the Ethernet to a remote "receiver site" where the signals were used to actuate and control the STIFF-FLOP arm. In some of the experiments, force

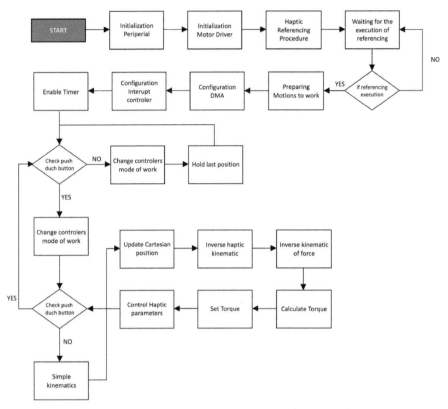

Figure 16.8   *RobinHand* manipulator and main program loop.

and tactile sensor information was collected at the robot arm–environment interface and fed back to the master system providing the user with appropriate resistance to their input movements when using the user input console, achieving tactile feedback.

16.4.1 Telemanipulation FCSD-UoS RobinHand H

The main specifications of this experimental study can be summarized as follows:

- Access to a private network can be obtained by using VPN server and FCSD infrastructure provided by a public network such as the Internet (University of Surrey, Dr. Tao Geng).

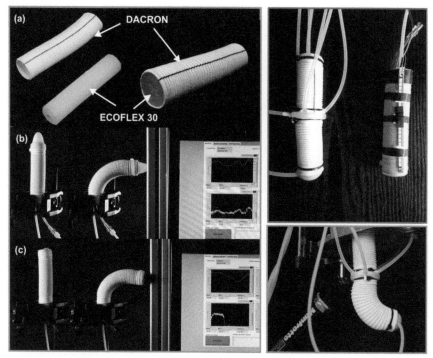

Figure 16.9   The FCSD flexible manipulator inspired by STIFF-FLOP; the construction and the first test of manipulator with pressure and force sensors.

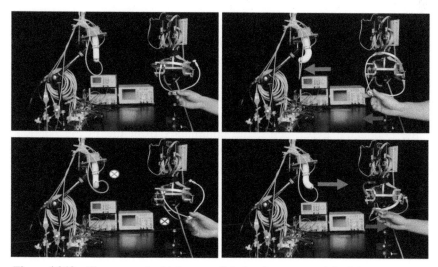

Figure 16.10   The pneumatic driving test of the haptic system and flexible manipulator.

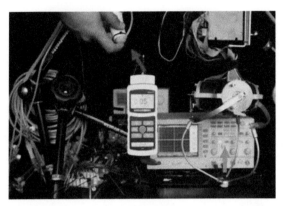

Figure 16.11   The feedback evaluation test of the haptic system and flexible manipulator.

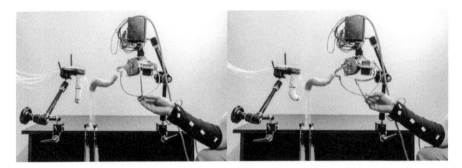

Figure 16.12   The full haptic – flexible manipulator – vibration sleeve test of feedback.

- A computer from FCSD operates as ROS_MASTER and publishes Delta coordinates X, Y, Z via the ROS topic, "/ sf_delta_pos." After that, the computer receives the published coordinates from Surrey University and converts those into the movement of the STIFF-FLOP arm (Figure 16.13).

In the conducted experiments, the commands from the FCSD haptic system FCSD Delta were sent to the robot controller at UoS actuating the pneumatic actuators of one STIFF-FLOP arm module. The test was conducted successfully and represents an important step in the development cycle of STIFF-FLOP. As shown in Figure 16.13 and on the video (available at the repository), when the Delta Haptic at FCSD is moving, the robot arm at UoS moves accordingly.

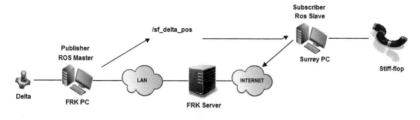

Figure 16.13 FCSD Haptic *RobinHand H* connection to the ROS system; performing tele-manipulation between FCSD (at left) and UoS (at right).

16.4.2 Telemanipulation FCSD-PIAP RobinHand F

The first testing of the STIFF-FLOP control feedback console was carried out via remote connection between FCSD (Zabrze) and PIAP (Warsaw). Figure 16.14 shows the Haptic *RobinHand F* manipulator during the connection, prepared for the STIFF-FLOP robot – in this case, the console was equipped with a force feedback mechanism. The console's design is based on the keypad Delta (i.e., parallel kinematics). The Delta man–machine interface allows three DOF positions in space.

It is possible to increase the number of degrees of freedom to a maximum of seven DOF. In order for this device to work as described, forward and

Figure 16.14 Remote telecontrol of STIFF-FLOP arm during FCSD-PIAP connection via the Internet (VPN).

inverse kinematic models are required. The main haptic algorithm enables users to touch, feel, and manipulate STIFF-FLOP modules through the force-feedback Delta device. The Delta parallel manipulator has large force reflection and high stiffness due to its parallelogram-type structure. Our Delta Haptic system was equipped with an electronic interface and software libraries, allowing the control to be conducted in the general ROS environment used in the framework of the STIFF-FLOP project. The communications interface between the Delta Haptic system and ROS was coded using a decimal format: XXX.X [mm], to achieve positions in the X, Y, Z space. Through these tests conducted between FCSD and PIAP, we could show that the connectivity between console and STIFF-FLOP arm prototype via the Internet works satisfactory. The test was conducted successfully and represents an important step in the development of STIFF-FLOP haptic control interface.

During the workshop carried out in Warsaw, we had the opportunity to integrate components of the control system. First, the preparation of all components was started by individual teams. FCSD successfully demonstrated the working of the Delta Haptic system, visualizing the motion of the user input in a 3D virtual environment. The next step was a preliminary test of the control console, taking into account the haptic feedback and evaluating the feedback strength. The test was carried involving a STIFF-FLOP manipulator prototype (Chapter 2), for improved actuation and force/torque sensing. Due to lack of properly working force sensors, testing of haptic force feedback had to be abandoned. In these experiments, the haptics were interfaced only with the ROS system, and the transmission of position from the haptics device to ROS was checked (Figure 16.15).

Additionally, a functional verification of the 2:1 scaled phantom model in frontal plane of the abdominal area (described in Chapter 17) for new STIFF-FLOP manipulator was carried out (Figure 16.16). This test provided a means to check every sensor placed on the sacrum below the urinary bladder and the colon (near the anus). Another three sensors were placed inside the urinary bladder and iliac vessels to measure the force acting on the wall of this body part.

16.4.3 Telemanipulation FCSD-KCL RobinHand F

During the project evaluation studies, the STIFF-FLOP arm was remotely controlled using the FCSD Delta. The STIFF-FLOP manipulator was placed above the up-scaled FCSD phantom model, allowing the robot tip to operate in the frontal plane of the abdominal area (Figure 16.17).

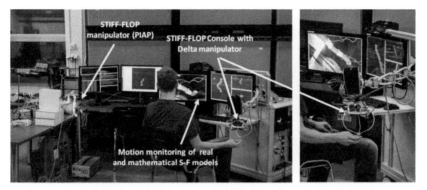

Figure 16.15   The STIFF-FLOP manipulator local control test.

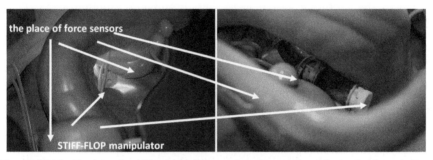

Figure 16.16   Functional verification of 2:1 scaled phantom models in the frontal plane (of abdominal area) for the new STIFF-FLOP manipulator (PIAP).

Figure 16.17   Telemanipulation evaluation studies at FCSD.

16.5 Integrating the Haptic Device RobinHand L with STIFF-FLOP Console

The final version of the *RobinHand L* actuator with seven DOF was adapted for STIFF-FLOP control console presented during the meeting in Torino (Figure 16.18). The first tests of the software integrating the actuator, the

Figure 16.18   Presentation and tests of the integrated system.

console, and the vibrating sleeve were very promising. Another stage of development was presented during the meeting in London where the console was modified in order to make it more user friendly. Modifications were also made to the wiring, electronic components, and control software – to scale the forces impacting the operator. Additionally, the electronic part was equipped with cooling and ventilation systems to prevent it from overheating. The full integration took place in London where the project was successfully completed (Figure 16.19).

The actuator was also adapted for the brand-new version of the console where the mechanical part was separated from the electronic module. Haptic *RobinHand L* was modernized to meet the needs of INCITE project [15, 16]. Design and ergonomics were improved. The most recent Robin STIFF-FLOP console is shown in Figure 16.20.

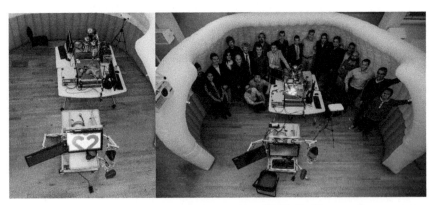

Figure 16.19   Final presentation of the STIFF-FLOP system.

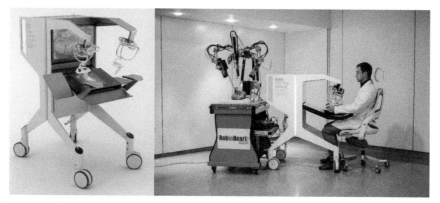

Figure 16.20 Robin STIFF-FLOP console: CAD model and integration with Robin Heart [16].

16.6 Conclusion

Biocybernetics Laboratory of the Heart Prosthesis Institute at the Cardiac Surgery Development Foundation is carrying out a Robin Heart medical robot project. Several models of the robot have been designed, made, and tested. The construction design and the control system, including the ergonomic console and actuator/motion controller, are continuously improved. In this regard, the greatest achievements of the Polish team include Robin Heart Shell 1 console, Robin Heart Shell 2 console, and Robin STIFF-FLOP console – with fully haptic motion actuator/motion controller and force feedback.

Currently, further testing of *RobinHand* haptic system is being conducted. All the results obtained so far indicate that it can meet customers' expectations and be put on the market. We hope that the knowledge and experience we have gained will allow us to provide surgeons with a comfortable workstation making remote operations safe and precise.

Acknowledgments

The project of flexible tool was supported in part by the European Commission within the STIFF-FLOP FP7 European project FP7/ICT-2011-7-287728. The authors thank K. Rohr and M. Jakubowski from the Prof. Z. Religa Foundation of Cardiac Surgery Development for their assistance with the technical aspects of the study.

References

[1] Xin, H., Zelek, J. S., and Carnahan, H. (2006). "Laparoscopic surgery, perceptual limitations and force: A review," in *Proceedings of the First Canadian Student Conference on Biomedical Computing*, Kingston, ON, 144.

[2] Bicchi, A., Peshkin, M., and Colgate, J. E. (2008). "Safety for physical human–robot interaction," in *Springer Handbook of Robotics*, eds B. Siciliano, O. Khatib (Berlin: Springer-Verlag), 1335–1348.

[3] Wurdemann, H. A., Secco, E. L., Nanayakkara, T., Althoefer, K., Lis, K., Mucha, Ł., et al. (2013). "Mapping Tactile Information of a Soft Manipulator to a Haptic Sleeve in RMIS," in *Proceedings of the 3rd Joint Workshop on New Technologies for Computer Robot Assisted Surgery*, Verona.

[4] Mucha, Ł. (2015). Interfejs użytkownika robota – przegląd urządzeń zadawania ruchu systemów sterowania telemanipulatorów. *Med. Robot. Rep.* 4, 39–48.

[5] Mucha, Ł., Lehrich, K., Nawrat, Z., Rohr, K., Lis, K., Sadowski, W., et al. (2015). Postępy budowy specjalnych interfejsów operatora robota chirurgicznego Robin Heart. *Med. Robot. Rep.* 4, 49–55.

[6] Katsura, S., Iida, W., and Ohnishi, K. (2005). Medical mechatronics— An application to haptic forceps. *Ann. Rev. Control* 29, 237–245.

[7] Fraś, J., Tabaka, S., and Czarnowski, J. (2016). "Visual marker based shape recognition system for continuum manipulators," in *Challenges in Automation. Robotics and Measurement Techniques*, eds R. Szewczyk, C. Zieliñski, M. Kaliczyñska (Cham: Springer).

[8] Nawrat, Z., Kostka, P., Lis, K., Rohr, K., Mucha, Ł., Lerich, K., et al. (2013). Interfejs operatora robota chirurgicznego – oryginalne rozwiązania sprzężenia informacyjnegoi decyzyjnego. *Med. Robot. Rep.* 2, 12–21.

[9] Gosselin, F., Martins, J. P., Bidard, C., Andriot, C., and Brisset, J. (2005). "Design of a new parallel haptic device for desktop applications," in *Proceedings of the IEEE Conference and Symposium on Haptic Interfaces for Virtual Environment*, 189–194.

[10] Mucha, L., Lis, K., Rohr, K., and Nawrat, Z. (2016). *Manipulator of a Medical Device with Auxiliary Motor and Encoder*. USA patent, US 9393688, FCSD.

[11] Lehrich, K., Lis, K., Nawrat, Z., Mucha, Ł., and Rohr, K. (2016). The application of 3D printing to the construction of medical manipulators prototypes. *Mechanik* 3, 224–225.

[12] Mucha, L., Lis, K., Rohr, K., and Nawrat, Z. (2017). *Manipulator of a Medical Device*. European patent, EP 2990005, FCSD.

[13] Lis, K., Mucha, Ł., Lehrich, K., and Nawrat, Z. (2015). *Steering Handle for a Motion Manipulator*. Utility models Poland, W.124541, FCSD.

[14] Mucha, Ł., Nawrat, Z., Lis, K., Lehrich, K., Rohr, K., Fürjesb, P., et al. (2016). Force feedback control system dedicated for Robin Heart Pelikan. *Acta Bio-Optica et Inform. Inżyn. Biomed.* Vol. 22, 146–153.

[15] Nawrat, Z., Förjes, P., Mucha, Ł., Radó, J., Lis, K., Dučsõ, C., et al. (2016). Force Feedback Control System Dedicated for Robin Heart Surgical Robot. *Procedia Eng.* 168, 185–188.

[16] Rohr, K., Fuïjes, P., Mucha, Ł., Radó, J., Lis, K., Dučsõ, C., et al. (2015). Robin Heart force feedback/control system based on INCITE sensors – preliminary study. *Med. Robot. Rep.* 4, 10–17.

PART V

Benchmarking Platform
for STIFF-FLOP Validation

17

Benchmarking for Surgery Simulators

Zbigniew Malota, Zbigniew Nawrat and Wojciech Sadowski

Prof. Z. Religa Foundation of Cardiac Surgery Development, Zabrze, Poland

Abstract

An increasing number of robotic surgery simulators can be used for validation study of surgical education curricula, their functionality, and testing of their proficiency. These simulators are equipped with a multilevel curriculum, designed with different levels of difficulty for effectively advancing robotic surgery abilities. The most important factor for obtaining an appropriate surgical simulation is creating quasi-natural geometry of surgical scene and physical characteristics of the used materials. In acquisition of classical endoscopic surgery skills, many of the initial challenges are related to a loss of depth perception, the fulcrum effect, and the use of new, different instruments.

Constructing simulators for a completely new type of tool, with variable capacity and controlled geometry, the authors faced a new challenge. The positions created were both research devices of new tools (for engineers) as well as trainers (for surgeons) to discover the optimal use of new functional features of tools for various types of operations.

To support the educational process, the virtual operating room for planning the surgery and training station has been prepared. In this article, the authors show the process of producing the surgical training stations and few examples of the latest realized specialized devices. This platform allows a geometric modeling of the body anatomy, but also the modeling of the physical properties of the living tissues.

Designed and implemented by the Foundation of Cardiac Surgery Development Biocybernetics Laboratory team, special research station modeling selected surgical scenes were used to study all versions of tools and a fully functional robotic Stiff-Flop surgical system.

17.1 Introduction

Minimally invasive surgical procedures are very complex motion sequences that require a high level of preparation and surgical skills training. New tools developed for the use of new medical procedures also require early testing. Benchmarking is an essential part of the design of prototypes. There are many types of simulators that are available for surgical skill training and device testing. Simulators can be broken down into two different groups: high-fidelity and low-fidelity. These models vary widely with respect to their level of fidelity or realism, compared with a living human patient. The fidelity of a simulator is determined by the extent to which it provides realism through characteristics such as visual cues, tactile features, feedback capabilities, and interaction with the trainee. There are a variety of simulators; the following can be a way to categorize them [1, 2]:

- Synthetic models and box trainers;
- Live animal models;
- Cadaveric models;
- *Ex vivo* animal tissue models;
- Virtual reality (computer-based) models;
- Hybrid simulators;
- Procedure-specific trainers;
- Robotic simulators.

Synthetic model trainers using physical objects usually involve models of plastic, rubber, silicone, and latex. These objects are used to render different organs and pathologies and allow a trainee to perform specific tasks and procedures [3]. A box trainer uses the clinically available instruments and optical system to manipulate "synthetic" tissues. Some physical simulators may also reproduce the feedback from the surgical environment. Artificial materials can effectively replace the natural bodies (anatomic sections or tissues) from euthanized animals and may provide approximate haptic feedback.

In general, benchmarking platforms are based on simulators which describe the anatomy, in particular the geometry of the structures involved in a surgical intervention. These platforms allow geometric modeling of the body anatomy, but also the modeling of the physical properties of the living tissues. The implementation of biomechanical properties is necessary to allow realistic interactions between surgical instruments and soft tissues, including deformations and cutting.

Surgery simulators can be classified into three categories, as follows [4]:

- First-generation simulators describe only the anatomy, in particular the geometry of the structures involved in a surgical intervention.
- Second-generation simulators additionally include the geometric modeling of the physical properties of the living tissues to allow realistic interactions between surgical instruments and soft tissues.
- Third-generation simulators combine anatomical, physical, and physiological modeling, for modeling some organic systems' function such as the cardiovascular, respiratory, or digestive systems.

The modeling of biological tissues for second- and third-generation simulators is very difficult. The biological soft tissues have nonlinear force-deformation properties and show viscous behavior. The properties of soft tissues are often anisotropic and heterogeneous and show hysteresis, relaxation, and creep behaviors. Additionally, these properties strongly depend on many factors, including temperature, pressure, and patient health. Dissected tissue often changes its mechanical properties, so literature data may greatly differ among themselves. It should also be noted that the shape and mechanical properties of animal bodies also significantly differ from human organs.

The Foundation of Cardiac Surgery Development is a well-known research center for surgical robotics. The development Robin Heart robot and mechatronic tools are underway for clinical application. The STIFF-FLOP project focused on a new kind of robot, design bioinspired by octopus anatomy. The benchmarking system equipped with a sensorized phantom of chosen surgical scene allowed to assess the progress at each stage of the design of these innovative surgeon's tools under the STIFF_FLOP project.

17.2 Testing and Training Station Description

17.2.1 The New Scaled Surgery Benchmarking Platforms

The Foundation of Cardiac Surgery Development (FCSD) prepared a surgery benchmarking platform for the STIFF-FLOP robotics tools' test and modeling of chosen surgical procedure. Based on some minimally invasive procedures, essential benchmarking scenarios have been defined and designed, and fabrication of special test rigs and objects (such as phantoms representing organs with variable stiffness) has been carried out.

Figure 17.1 presents the different modules being connected with a central computer control and a monitoring system. This system is also developed for

Figure 17.1   STIFF-FLOP robot testing system diagram.

supporting educational activities, and provides, therefore, several components enabling the control of various elements (such as the electromechanical, electro-pneumatic, or electro-hydraulic modules). In addition, the virtual module is envisioned to provide to the students a virtualized version of the whole system (even though no interaction mean with the operation site is currently implemented).

To model the functions of some organic systems such as the cardiovascular or digestive systems, it is necessary to take into account anatomical, physical, and physiological properties. There is an additional degree of complexity due to the coupled nature of physiological and physical properties. Two types of test stands are designed and manufactured:

- Benchmarking platform – mainly reflecting functional characteristics for medical procedures including the obstacle track for training and evaluation of surgical skills;
- Anatomical – mainly based on the anatomical model which reflects the real geometry of the bodies.

Due to the heterogeneity of the biological material, the modeling of mechanical properties of human organs by artificial organs' material is very difficult. The selected artificial materials cover the basic range of variability of the mechanical properties of the bodies used to simulate the surgical procedures. Our platform fulfills the conditions of second-generation simulators as it describes the anatomy, in particular the geometry of the structures involved in a surgical intervention, and it includes the modeling of the physical properties of the living tissues. The introduction of biomechanical properties to our platforms is essential to allow realistic interactions between surgical instruments and soft tissues, including deformations during basic manual operations like cutting or sewing. Due to the large size of the STIFF-FLOP arm prototype at the time, it was necessary to make the platforms in 2:1 scale.

The benchmarking platform (Figure17.2), like the anatomical model, includes a flexible abdominal wall made from silicone. It is fixed by special

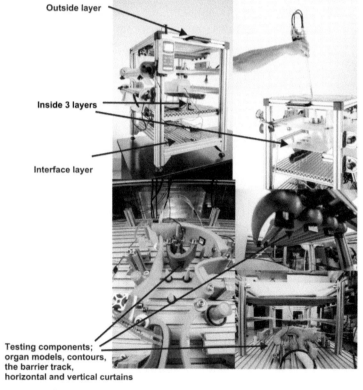

Figure 17.2 The multilayer benchmarking platform [5].

couplings to the chassis so that it can be freely configured (move and change the number of layers). However, in contrast to the anatomical model, the test platform enables research in 3D space, and is freely scalable (maximum scale 4:1). The frame of the platform and basis allows attachment of any organ model and sensors for testing. In addition, we can change the workspace by the movement of profiles in X, Y, and Z directions.

The operational area allows installation of flexible elements to simulate abdominal organs, and can contain measuring sensors. Each element can be easily adapted to the needs of the benchmark. The original solution to our benchmarking platform is the usage of the so-called multi-story system allowing the free distribution of platform elements in 3D space. This system consists of a number (one to three) of additional flexible planes, stretched at different heights and different widths. These planes can play functionality features of a variety of abdominal organs by attaching the various testing components and also properly formed organs of different shapes by cutting (incision) of different shapes. Sensors may serve as additional test elements on their own. In this way, an additional 3D plane (contours, the barrier track, and horizontal and vertical curtains) makes easy modeling of any surgical procedures.

17.2.2 Sensorized Operation Site

Technical benchmarking of the robotic arm performance will be realized by measurements of the position of the end tools and the force acting on the soft tissues. This apparatus is presented in Figures 17.3 and 17.4. The design of this system is modular to enable an easy reconfiguration of the setup on demand. Flexible elements installed on the apparatus permit emulating the human cavity.

As illustrated in Figure 17.3, the sensors that were envisioned and embedded were as follows:

- Flexi Force Sensor PHI-3100_0 with FlexiForce Adapter–1120
- Foil electrical resistance sensor Tz Fs-10/350 (Tenmex, Poland)
- Force Sensors Mark-10 (Mark-10 Corporation, United States)
- Plastic Fiber Optic Sensor– D10 Expert Sensor (Banner, United States)
- Simple displacement sensors (resistor) and push-buttons.

During the Surgical STIFF-FLOP Workshop in Zabrze (Figure 17.5), the current advancement of these systems has been presented, and discussed with the partners, surgeons, and medical students.

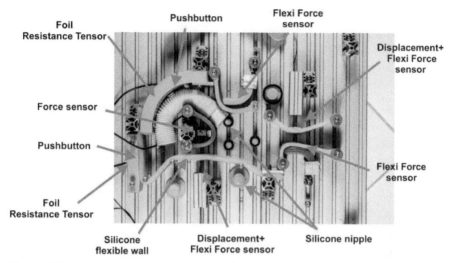

Figure 17.3 Illustration of the instrumentation of the operational site. The figure illustrates the current system that can be extended with some of the systems proposed within this section.

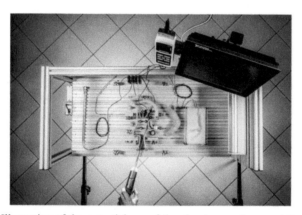

Figure 17.4 Illustration of the potential use of the phantom environment for capturing and analyzing the action of the robotic arm. Experiments can use the STIFF-FLOP arm directly inside such a system.

The results obtained during the workshop have been used to improve the next version of this prototype (Figure 17.6).

Through a workshop with surgeon partners, the following decisions were made to improve the system (Figure 17.6):

Figure 17.5 Illustration of some evaluation platforms that could be extended to perform some training of the STIFF-FLOP arm with surgeons and/or students: (a) presentation of the platforms during the Zabrze workshop, (b) Robin Heart control system using vision, (c) laparoscopy training stations requesting to perform some specific tasks, and (d) virtual laparoscopy training session [6].

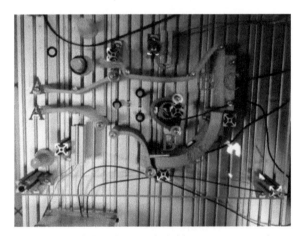

Figure 17.6 First analysis of the phantom hybrid system: suggestions provided by the surgeons to improve the current setup during the Zabrze workshop.

- The fixed points at the entry of the phantom (A) should be replaced with non-rigid structures with embedded sensors in order to understand the lateral pressures realized as well as the traction forces.
- More vertical nipples should be placed (B) in order to make specific exercises of moving rubber circles from one pin to the other.
- These new pins should be placed close to the walls annotated (C). In each of these walls, a pressure sensor should be placed to verify eventual wrong movement.

It was concluded that due to the current module prototype module size, such platform should be realized at a scale of two or three times the current size. In addition, the esophageal model illustrated in Figure 17.7 was prepared.

17.2.3 The Scaled Surgery Benchmarking Platforms

Based on selected minimally invasive procedures (as proposed by STIFF-FLOP medical partners), FCSD defined and built essential benchmarking scenarios and designed and fabricated special test rigs and phantoms representing organs with variable stiffness.

The phantoms combine anatomical, physical, and physiological modeling of the functions of natural organs systems.

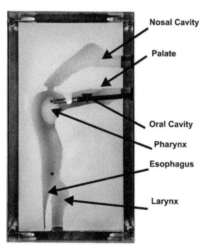

Figure 17.7 Anatomical phantom for otolaryngology. One objective is to study the integration of all the sensing elements within some elastic elements made of silicone and urethane rubber that would enable the provision of a realistic emulation of human cavities.

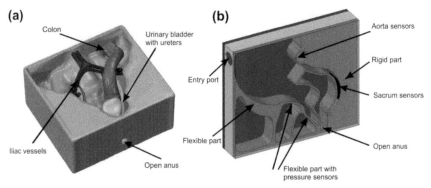

Figure 17.8   The design of a new anatomical phantom model of the lower gastrointestinal tract (scale 2:1) in the frontal (a) and sagittal (b) plane.

To simulate surgical procedures in the quasi-3D platforms (in scale 2:1) of the lower gastrointestinal tract (Figure17.8), a model based on the anatomical shape of the gastrointestinal tract in the sagittal and frontal plane was created.

While building platforms, the authors paid special attention to the internal geometry of the operating field and functionality of minimally invasive procedures. The generally available materials like silicone and urethane rubber (Smooth-On, Inc., United States) with different mechanical properties have been used.

Due to the heterogeneity of the biological material, the modeling of mechanical properties of human organs using artificial organ material is very difficult. Selected artificial materials that cover the basic range of variability of the mechanical properties of the human organs tissue to simulate the selected surgical procedures have been used. The platform satisfies the requirements which were proposed for this second-generation simulator because it describes the anatomy, in particular the geometry, of the structures involved in a surgical intervention, and includes the modeling of the physical properties of the living tissue. The introduction of biomechanical properties to our platforms is essential to allow realistic interactions between surgical instruments and soft tissue organs, including deformations during basic manual operations like grasping, retraction, cutting, or suturing.

Due to the large size of the STIFF-FLOP robot arm prototype at the time, it was necessary to realize the test platforms at a scale of 2:1. It is noted that the quasi-2D phantom models of the lower gastrointestinal tract were made only in two planes: sagittal and frontal.

The anatomical areas of this model allow studying and benchmarking surgical robot systems, as follows:

- Colonoscopy or sigmoidoscopy – for the endoscopic examination of the large bowel and the distal part of the small bowel with a camera passed through the anus for visual diagnosis and for biopsy or removal of suspected colorectal cancer lesions;
- Proctocolectomy – for the surgical removal of the rectum and all or part of the colon;
- Colectomy – for the surgical resection of any extent of the large intestine (colon resection).

The phantom model includes a flexible abdominal wall made from silicone (Figure 17.9). The operational area allows the installation of flexible elements to simulate abdominal soft-tissue organs, and has been equipped with a suite of sensors, including force and tactile sensors to reach the functionality needed for minimally invasive test procedures.

The new (modernized) phantom model in the sagittal plane included (Figure 17.10) the following:

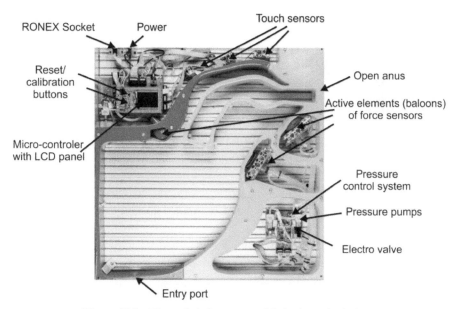

Figure 17.9 The scaled phantom models in the sagittal plane.

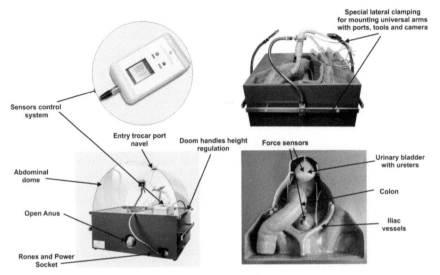

Figure 17.10   The scaled phantom models in the frontal plane.

- Reduced-width silicone walls;
- Replacement of pressure sensor and force sensors (originally potentiometer-based sensors) with new highly sensitive force sensors;
- New electronics;
- New calibrated sensors;
- The possibility to change the mechanical properties of the wall;

The new force sensor was based on the Honeywell pressure sensor and pneumatic system with a balloon as the active element. The anatomical shape with the abdominal surface was made from PET (polyethylene terephthalate, P.W. Masterchem S.J., Poland) with different thicknesses and different numbers of layers (transparent or opaque).

This phantom model can be used in two versions, the first with dome containing one Trocar port and the second without a dome but with special lateral clamping for mounting universal arms with ports, tools, and cameras. The flexible artificial organs like colon, urinary bladder with ureters, iliac vessels, and anus were made from silicone and urethane rubber (Smooth-On, Inc.) through molding.

This phantom model was equipped with pressure sensors based on the Honeywell 1PSI AXIAL sensors, similar to the sensors used in the phantom model of the sagittal plane. However, instead of adjustable balloons as the active elements, the sensor elements that are inflated to a fixed level are used. The sensors (for the measurement of the strength of up to 4 N with an accuracy of 0.1 N) were placed below the bladder, on the sacrum, iliac vessels, aorta, and the colon near the anus.

17.2.4 The Virtual Reality Model

Designing physical models of test devices should be preceded by projects in the virtual space.

The 3D virtual-reality technology can verify the basic functional assumptions at the design stage of simulators, as well as possibly enhance surgeons' learning experiences by providing them with a heuristic and highly interactive simulated virtual environment. The created virtual models are independent test objects that can be used to plan surgical operations or training strategies (Figure 17.11).

In the design process, virtual phantom models were used to verify and check the functionality of both the flexible STIFF-FLOP arm and the haptic control system with haptic feedback.

Virtual reality technology is an interdisciplinary technology, integrating CAD/CAM technology, artificial intelligence, computer networks, and sensor technology. It is widely used in the design and testing of mechanical models. EON Studio is the software tool using graphical interfaces and used for research and development of real-time 3D-modeling applications. This method has been used by FCSD for interactive virtual modeling of the flexible STIFF-FLOP robot arm and motion simulation with interaction between the model of the STIFF-FLOP arm and the surgical environment. The virtual scene reflects the real phantom model of the abdomen in the frontal plane with elements of flexible organs like colon, urinary bladder with ureters, and iliac vessels. Using virtual reality technology to plan surgery procedures

Figure 17.11   The virtual reality phantom models.

increases the efficacy of methods and helps to verify the design and concept of STIFF-FLOP arm.

17.3 Conclusion

An increasing number of robotic surgery simulators can be used for the validation study of surgical education curricula, their functionality, and testing medical procedures.

The surgical simulator must not only accurately maps the anatomical details and deformation of the organ, but also feed-back realistic tool-tissue interaction forces. Therefore, development of realistic surgical simulation systems requires accurate modeling of organs/tissues and their interactions with the surgical tools.

However, to the heterogeneity of the biological material, the modeling of mechanical properties of human organs by artificial organs material is very difficult. Artificial materials selected by us cover the basic range of variability of the mechanical properties of the bodies used to simulate the surgical procedures.

The artificial surgical scene and described devices for testing tools and surgeons create possibility of standardization for the educational and research process. Thanks to various artificial surgical scenes, we can better and more effectively assess the usefulness of new surgical instruments (mechanical, mechatronic, and robotic) for use in various medical procedures.

Acknowledgments

The project of flexible tool was supported in part by the European Commission within the STIFF-FLOP FP7 European project FP7/ICT-2011-7-287728. The authors thank M. Jakubowski and A. Klisowski from the Prof. Z. Religa Foundation of Cardiac Surgery Development for their assistance with the technical aspects of the study.

References

[1] Sturm, L., et al. (2007). *Surgical Simulation for Training: Skills Transfer to the Operating Room*. ASERNIP-S Report, No. 61. Adelaide, SA: ASERNIP-S.

[2] Tsuda, S., and Scott, D. (2009). Surgical skills training and simulation. *Curr. Probl. Surg.* 46, 271–370.

[3] Cisler, J., and Cisler, J. (2006). Logistical Considerations for Endoscopy Simulators. *Gastrointest. Endosc. Clin. N. Am.* 16, 565–575.

[4] Delingette, H., and Ayache, N. (2005). Hepatic surgery simulation. *Comun. ACM*, 48, 31–36.

[5] Malota, Z., Nawrat, Z. and Sadowski, W. (2014). Minimal invasive surgery simulation (symulatory chirurgii małoinwazyjnej). *Eng. Biomat.* 17, 2–11.

[6] Malota, Z., Nawrat, Z., Kostka, P., Sadowski, W., Rybka, P., and Rohr, K. (2013). "Stanowiska treningowo-badawcze narzêdzi i robotów chirurgicznych Fundacji Rozwoju Kardiochirurgii im. prof. Z. Religi w Zabrzu," in *Proceedings of the Advances in medical robotics (Postêpy robotyki medycznej). Red.: L. Leniowska, Z. Nawrat*, Rezszow, 151–163.

18

Miniaturized Version of the STIFF-FLOP Manipulator for Cadaver Tests

Giada Gerboni, Margherita Brancadoro, Alessandro Diodato, Matteo Cianchetti and Arianna Menciassi

The BioRobotics Institute, Scuola Superiore Sant'Anna,
Pontedera (PI), Italy

Abstract

The previous version of the STIFF-FLOP manipulator (reported in Chapters 2 and 3) has demonstrated to have all the capabilities to perform an effective surgical operation, being able to bend in any direction, to sense the external forces, and to elongate and to change its stiffness. On the other hand, its size is not yet compatible with standard tools used for minimally invasive surgery (i.e., standard trocar ports). The work described in this chapter has focused on the development of a miniaturized version of the STIFF-FLOP manipulator, with limited sensing and stiffening abilities. This manipulator has been equipped with a camera and used in a realistic scenario, such as cadaver tests. The STIFF-FLOP robotic optic arm seemed to acquire superior angles of vision of the surgical field and neither intraoperative complications nor technical failures were recorded.

18.1 Requirements for Manipulator Usability in Cadaver Tests

The miniaturized version of the soft manipulator has been designed taking advantage of the expertise gained in developing previous STIFF-FLOP manipulator versions. Modularity, novel actuation technologies, and smart geometries have been exploited to match the following specific requirements:

- The entire device has to fit the maximum size of traditional trocars, and thus a diameter smaller than 15 mm is strictly required.

- The manipulator should allow adjusting the position and direction of the end-effector separately, by keeping the shaft stable; thus more than four degrees of freedom (DOFs) are necessary to enable the required flexibility. This means that at least two modules, each one integrating a 4-DOF flexible fluidic actuator (FFA), are required.
- The employed FFA must not generate free elastomeric deformation phenomena (e.g., lateral chamber expansions, ballooning, etc.) during its use. This would ensure a reduced risk of breakage and an efficient use of fluidic actuation.
- The compliance of the soft modules must be preserved as much as possible in order to squeeze out the deflated manipulator from tight spaces in case of emergency.
- A free lumen of about 4–5 mm is required for feeding through surgical tools (e.g., laparoscopic graspers, RF tools, etc.), or for allocating the wires of a laparoscopic camera for vision, which can be fixed at the tip.
- The manipulator maneuverability must take into account the effect of downline loads, including the weight of additional modules.
- The manipulator has to be able to exert forces and handle unwanted contacts while maintaining the distal extremity (where the guided surgical tool is located) at the required position for accomplishing the surgical task.

For the STIFF-FLOP arm version used for cadaver tests, the efforts have been mainly focused on replicating the mobility performances already obtained with the previous large versions of the STIFF-FLOP arms, but at a smaller scale.

18.2 System Adaptation

Figure 18.1 shows the design of the whole miniaturized STIFF-FLOP manipulator system, which is composed of two pneumatically actuated hollow modules. The manipulator is attached to a rigid shaft, which is also hollow and serves as support during the MIS procedure. This rigid shaft connects the flexible manipulator to the pressure-control system and it is necessary to rigidly interact with the fulcrum at the entry port where the trocar is placed. The shaft can be controlled either manually by the medical personnel or by a robot.

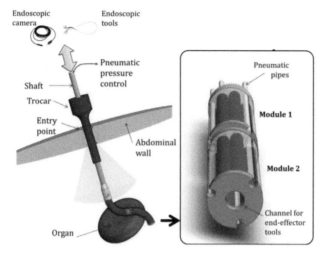

Figure 18.1 (Left) Envisaged system applied to a MIS scenario. (Right) Manipulator design with two soft modules able to bend in every direction and elongate (4 DOF).

The two modules, as specified in Figure 18.1 – right, are both able to perform omnidirectional bending and elongation motion in response to the level of pressure applied to their inner chambers. A thin and hollow connection merges the two modules while keeping the chambers of the two modules aligned.

The first module is rigidly connected to the shaft in which the six fluidic pipes are guided to reach the external pressure control system. The pipes from the second module are guided along the outer walls of the first module not to occupy the inner lumen, which must be preserved for feeding surgical instruments or camera wires. Then all the pipes of all the modules are guided through the hollow shaft connecting the chambers to the external pressure control system.

Figure 18.2 shows a detailed view of a single module, which is 50 mm in length and 14.5 mm in external diameter, making it suitable for MIS applications employing standard trocars. The module contains three pairs of pneumatic chambers measuring 3 mm in diameter, designed with the approach detailed in Chapter 3.

As visible in Figures 18.1 and 18.2, a module is made up of six cylinders. This is because for each DOF, a pair of chambers is engaged. This choice is due to the fact that the module has to preserve a reasonably large hollow lumen (4.5 mm in diameter) for feeding through instruments (or devices or

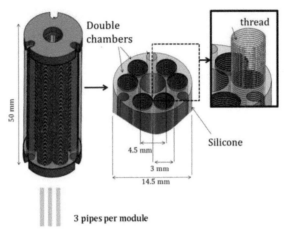

Figure 18.2   Single-module design with relevant quotes. The section view in the middle shows the inner arrangement of the chambers. The inset shows an exploded view of the thread used for avoiding the lateral chamber expansion.

stiffening mechanisms), and thus the residual space left in the module limits the maximum diameter of the chamber's circular cross section. Hence, in order to fill up the space available with cylindrical chambers, two cylinders of a smaller diameter are used in parallel for each DOF. An internal pipe connects the two chambers related to the same bending motion; so the pair of chambers can be considered as a single larger chamber, which is inflated by just one pipe from the central pressure-control system. Two end caps, made of a stiffer silicone (depicted in blue in Figure 18.2), are included at the top and bottom parts of the module, hermetically closing the chambers. The bottom of the module is designed to embed also the three pipes for the pneumatic pressure supply of each pair of cylindrical chambers.

In order to maintain the inner central lumen free, the module design includes three little external grooves in the silicone body for allocating the pipes coming from the distal module. In these grooves, the pipes can easily slide longitudinally while the module is working (bending and elongating). The grooves are shaped such that the pipes cannot escape from them. In case of more than two modules, a more efficient management of pipes coming from the distal modules should be identified.

The detailed manufacturing process of the small-sized modules of the STIFF-FLOP arm for cadaver tests follows the same strategy of the large-scale modules previously described in Chapter 3.

18.3 System Modeling and Characterization Methods

The miniaturized version of the manipulator intended for cadaver tests does not include an active system for stiffening (i.e., the granular jamming chambers). The elongation/bending actuation by itself produces some structure stiffening, which can be quantified. For this reason, the manipulator was modeled in order to characterize the stiffness of the single module when actuated [1].

Thanks to this module analysis, it is possible to predict the behavior of the entire manipulator when more modules are combined together: this can be important information for improving the design. Indeed, with such a model, the maneuverability limitations due to downline loads (i.e., the forces applied to the top of the distal module) can be estimated. Therefore, every module can be modeled in relation to the number of modules that follow it. In addition, the fluidic actuation efficiency of the soft module with only cylindrical elongating chambers' design can be evaluated.

Concerning the elongation motion primitive (eMP), the module can be seen as a spring, having constant K_L, which can elongate thanks to the force generated by the same pressure applied into all the six (2×3) chambers. Hence, the balance of the forces acting on the module during the eMP can be expressed as in Equation (18.1) (Figure 18.3 – left). The external loads are expressed by F_e.

The bending motion primitive (bMP) envisages a balance of the torque values around the base of the module. In this case, the module appears as a torsional spring, having constant K_t, which rotates by an angle θ when a single pair of chambers is inflated with pressure P. Considering the complexity of precisely modeling the module's actuation, a linear dependency of the resulting torque from the imposed chamber pressure was assumed. The balance of torque values, including the external torque (τ_e), is expressed as in Equation (18.2) (Figure 18.3).

Factor γ in Equation (18.1) represents the efficiency of the actuation. In every actuator, not all the input energy is totally converted into the output energy as part of this energy is dissipated through various phenomena (including friction, defects, leakages, and heat). Parameter α in Equation (18.2) is an unknown parameter, which takes into account the actuation efficiency and the chambers' contact area and arm with respect to the central axis.

In order to estimate the parameters of the model in Equations (18.1) and (18.2) (i.e., elastic constants K_L and K_t and factors γ and α), dedicated experimental tests have been carried out.

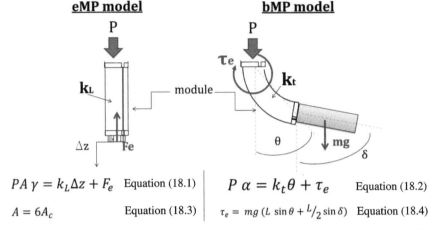

eMP model

bMP model

$PA\gamma = k_L\Delta z + F_e$ Equation (18.1) $P\alpha = k_t\theta + \tau_e$ Equation (18.2)

$A = 6A_c$ Equation (18.3) $\tau_e = mg\,(L\,\sin\theta + L/2\,\sin\delta)$ Equation (18.4)

Figure 18.3   (Left) Model of the module during eMP in response to the pressure P supplied to all the chambers. (Right) Model of the module during bMP in response to the pressure P supplied to a single couple of chambers in the proximal module. A_c is the effective area of a single chamber, m and L are, respectively, the mass and the length of a module, θ is the bending angle, and δ is the angle described by the extremities of "external load." The external load is represented by another module having the same dimensions and weight of the first and generates the torque τ_e.

For measuring the displacement of the modules, we used the tracking system Aurora (by NDI medical), provided with small probes (Aurora Mini 6 DOF Sensor 1.8 mm × 9 mm) positioned on key points of the manipulator.

For the estimation of the parameters of Equation (18.1), two data sets were collected, considering two different eMP conditions:

- **eMP_c1** The module can freely elongate in response to pressure P being supplied to all the chambers (Figure 18.4 – left).
- **eMP_c2** Isometric condition: The module is constrained to its initial length by positioning a ground-fixed load cell (ATI-Industrial Automation F/T sensor nano17) at its top (Figure 18.4 – right).

By applying Equation (18.1) to each of these conditions, the following system holds:

$$\begin{bmatrix} 0 \\ F_e \end{bmatrix} = \begin{bmatrix} PA & -\Delta z \\ PA & 0 \end{bmatrix} \begin{bmatrix} \gamma \\ k_L \end{bmatrix} \tag{18.5}$$

For the estimation of the parameters of Equation (18.2), the following two bending MPs (bMP) were considered:

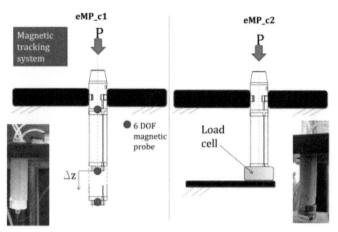

Figure 18.4 (Right) Bench test of the test evaluation of the displacement performed during free elongation motion (eMP_c1). (Left) Bench test of the test for measuring the force exerted by the module in the longitudinal direction on the load cell (eMP_c2).

- **bMP_c1** The module can freely bend in response to pressure P being supplied to a single pair of chambers (Figure 18.5 – left).
- **bMP_c2** The module can bend in response to pressure P being supplied to a single pair of chambers (Figure 18.5 – right). In this case, an external torque is applied. In a realistic scenario, this external torque would be generated by the presence of another module attached, for example, at the top of the first module (Figure 18.5 – right).

By applying Equation (18.2) to each of the above conditions, the following equation can be found:

$$\begin{bmatrix} 0 \\ \tau_e \end{bmatrix} = \begin{bmatrix} P & -\theta_1 \\ P & -\theta_2 \end{bmatrix} \begin{bmatrix} \alpha \\ k_t \end{bmatrix} \tag{18.6}$$

Equations (18.5) and (18.6) have the form:

$$Y = Ax \tag{18.7}$$

These two equations can be solved by using the pseudo-inverse $A\dagger$ approach (least squares method) for estimating the vector parameters.

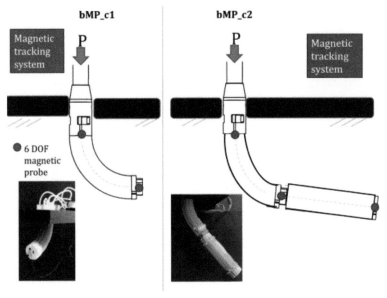

Figure 18.5   (Right) Bench test for the evaluation of the bending angle θ_1 performed during free bending motion (bMP_c1). (Left) Bench test for the evaluation of the bending angle θ_2 performed during loaded bending motion (bMP_c2). In this case, another module, provided with another six DOFs magnetic probe at the top, is used as load.

18.4 Results of Characterization

The results about the tests and model parameters estimation for eMP are summarized in Table 18.1. The module is able to elongate up to 27% during free body elongation, while the maximum force developed along the main axis was recorded to be 3.6 N before buckling phenomenon appears. Table 18.1 also reports the estimation reliability of the eMP model, expressed by Equation (18.1), in predicting the force developed by the module in terms of R^2 and mean and standard deviation of residual errors.

Table 18.1   Test results and model parameter estimation for eMP

Maximum elongation	13.67 mm
Maximum force exerted	3.6 N
γ	0.91
k_L	320.59 N/m
Mean $\pm$ std of residual errors	-0.09 ± 0.2 N
R^2	0.96

The same type of summary regarding the tests and model parameters estimation during bMP is reported in Table 18.2. The maximum angle achieved in unloaded condition (Figure 18.5, left) is 38.46°, while when the load is attached, it reaches 25.9° (Figure 18.5, right), and thus the angular difference between the two conditions is 12.56°. This difference is also reported for the entire data set in Figure 18.6. The estimation reliability of the bMP model is also reported in Table 18.2.

The models developed for the module in both eMP and bMP demonstrated to correctly predict the behavior of the real module. In a sense, this type of analysis allows to estimate the effective workspace of the manipulator, because it considers the effects of all the downline loads possibly applied to each module. In fact, Equations (18.1) and (18.2) could be recursively used for evaluating the force and the torque applied to the tip of each module. The modules are rigidly fixed one to another, and thus it is possible to assume

Table 18.2 Test results and model parameter estimation for bMP

Max free bending angle	38.46°
Max loaded bending angle	25.9°
α	8.15E^{-8} Nm/Pa
k_T	2.27E^{-4} Nm
Mean $\pm$ std of residual errors	-1.39E$^{-4}\pm3.84$E^{-4} Nm
R^2	0.85

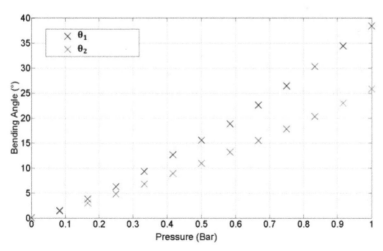

Figure 18.6 Bending angle recorded under the two conditions shown in Figure 18.5: unloaded (blue) and loaded state (orange).

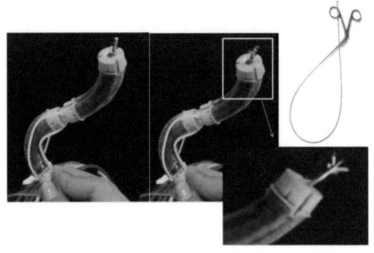

Figure 18.7 The soft manipulator is articulated while carrying a flexible endoscopic instrument (biopsy tool).

equilibrium of force and torque from the one module distal part to the base of the following module. However, two modules are already enough to equip the surgeon with the required flexibility to approach the target.

From Figure 18.6, it is possible to observe the effect of the load on the progression of the bending angle. Indeed, all the values of the free bending angle are bigger than the loaded angles. Furthermore, a relatively high force can be exerted by the module in the axial direction (Table 18.1).

The results about the actuation efficiency γ, reported in Table 18.1, confirm that the use of cylindrical chambers with thread, which do not expand but only elongate, improves the performance of the overall STIFF-FLOP robot system.

Finally, Figure 18.7 shows the manipulator provided with a commercially available flexible biopsy tool for standard endoscopes, demonstrating that the manipulator is able to articulate and withstand loads produced by the flexible instrument inside. The soft manipulator can be therefore used to approach difficult-to-reach targets during MIS procedures in a safe manner.

18.5 Prototype for Cadaver Test (with Integrated Camera)

The previously described two-module manipulator has been equipped with an endoscopic camera and used for tests in human bodies (cadaver tests).

Figure 18.8 shows this manipulator assembled for the cadaver tests and lists the specifications regarding the selected endoscopic camera (by Misumi Electronic Corp.).

In this case, even smaller fluidic pipes (size 1 mm of external diameter) were employed and passed into the lumen of the device. This was to facilitate the passage of the manipulator through the trocar port several times during the operation. In this way, the pipes are more protected and the modules showed an increased robustness.

The aluminum shaft, used for supporting the manipulator, has a diameter of 10 mm, thus allowing easy sliding of the shaft into the 15 mm trocar during the operation.

The entire device is firmly fixed to the operating table attached to an articulated arm (the Martin's Arm or Aurora arm) by means of a 3D printed connection, which also embeds the connection of the pneumatic pipes connecting air chambers with the remote control system and the camera USB connection. Figure 18.9 shows a panoramic view of the operating room, depicting the assembled device attached to the Martin's arm above the table.

Finally, the camera horizon is adjusted with respect to the upward movements of both the modules, once the manipulator is positioned in the appropriate operating configuration.

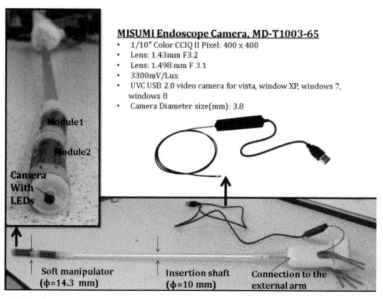

Figure 18.8   Soft two-module camera manipulator assembly and endoscopic camera details.

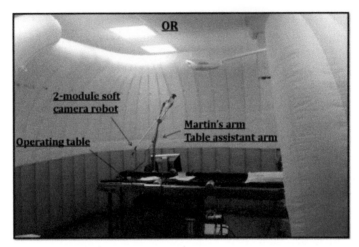

Figure 18.9   Panoramic view of the operating room (OR).

Figures 18.10 and 18.11 show pictures taken during the operations via laparoscopy on the human cadaver, performed at the Institute for Medical Science and Technology (IMSaT), University of Dundee, on October 13th, 2015 (carried out to appropriate ethical standards [2]).

The soft manipulator as well as other operative instruments and a supporting laparoscope have been inserted in the cadaver's inflated abdomen through a 15-mm trocar port. While one surgeon maneuvered the surgical

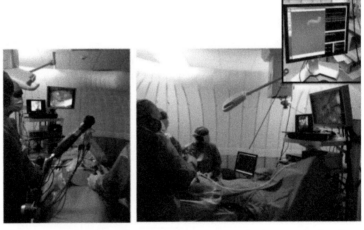

Figure 18.10   Surgeons controlling the soft camera manipulator and other laparoscopic devices.

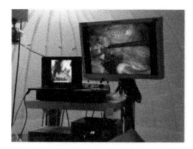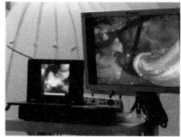

Figure 18.11   (Left) Vision of the target region through the soft camera module. (Right) Vision of the target region and soft manipulator through the laparoscope.

instruments, the other was using the joystick to control the STIFF-FLOP arm using the images from the robot's camera as feedback. Another surgeon guided a standard laparoscope to document the effective behavior of the soft manipulator underneath the abdominal wall (Figure 18.10). In Figure 18.11, the images from the endoscope (on the right screen) are shown and the images from the camera on the STIFF-FLOP manipulator (on the left screen) are also illustrated.

The STIFF-FLOP manipulator with camera confirmed its effectiveness in bringing a familiar endoscopic tool (as for example the camera) toward body areas which would not be accessible by standard straight and rigid laparoscopes. The soft and flexible nature of the manipulator demonstrated to be the key elements for reaching remote areas and overcome anatomical obstacles without damaging tissues and organs. Furthermore, thanks to the stable control of the pressure in the manipulator's pneumatic chambers, the surgeons were able to accomplish the relevant steps of the planned surgical procedure (the total mesorectal excision, TME).

Once the operation was complete, the camera was removed from the STIFF-FLOP arm prototype and the soft silicone elements were discarded.

References

[1] Abidi, H., Gerboni, G., Brancadoro, M., Fras, J., Diodato, A., and Cianchetti, M. et al. (2017). "Highly Dexterous 2-module Soft Robot for Intra-organ Navigation in Minimally Invasive Surgery", *Int. J. Med. Robot. Comput. Assist. Surg.* e1875.

[2] Arezzo, A., Mintz, Y., Allaix, M. E., et al. (2017). *Surg. Endosc.* 31:264.

19

Total Mesorectal Excision Using the STIFF-FLOP Soft and Flexible Robotic Arm in Cadaver Models

Marco Ettore Allaix[1], Marco Augusto Bonino[1], Simone Arolfo[1], Mario Morino[1], Yoav Mintz[2] and Alberto Arezzo[1]

[1]Department of Surgical Sciences, University of Torino, Torino, Italy
[2]Department of General Surgery, Hadassah Hebrew University Medical Center, Jerusalem, Israel

Abstract

In this chapter, we discuss the perspectives related to the use of a soft and flexible robotic camera during a series of rectal resections with total mesorectal excision (TME) that were performed in two human cadaver models. The robotic prototype comprised two modules that are 60 mm long and 14.3 mm large. The robot is connected to a rigid shaft that is in turn attached to an anthropomorphic robotic arm with six degrees of freedom. Briefly, three standard laparoscopic tools were employed to perform the surgical procedure. After the splenic flexure mobilization and the inferior mesenteric vessel division, the mesorectum was entirely *en bloc* excised. Neither intraoperative adverse effects nor technical issues were recorded. This preliminary experience shows that the STIFF-FLOP soft and flexible robotic optic arm is an effective tool that is characterized by a better vision of the surgical field than the standard laparoscopic rigid camera. The implementation of new soft and flexible robotic systems may help surgeons overcome the technical issues that are encountered during challenging minimally invasive surgical procedures performed using the standard rigid camera.

19.1 Introduction

While the laparoscopic approach to colon cancer has been widely adopted worldwide, with significantly better short-term outcomes than open surgery and similar oncologic outcomes [1–4], the current evidence about the role of laparoscopic resection for rectal cancer is controversial.

The introduction and rapid implementation of routine total mesorectal excision (TME) during resection of the mid and low rectum for cancer has led to a significant decrease in local recurrence rates [5] with subsequent improvement of long-term survival [6]. However, the open approach still represents the standard of care for the elective surgical treatment of rectal cancer, based on controversial results of recently published randomized controlled clinical trials. While the COLOR II trial showed that the minimally invasive approach is not inferior to the open approach, reporting similar short-term oncologic outcomes including resection margins and completeness of the mesorectal excision [7], and no significant differences in local recurrence rates and disease-free and overall survival at 3 years [8] between the two approaches, both the ACOSOG [9] and the ALaCaRT [10] randomized controlled trials (RCTs) failed to prove the non-inferiority of the minimally invasive approach regarding the pathology results, including completeness of TME and clearance of both radial and distal resection margin in stage 2 and 3 rectal cancer patients.

The current evidence from RCTs and prospective comparative studies [11, 12] shows that the laparoscopic approach to selected patients with both high and mid/lower resectable rectal cancer rectal resection has clinically measurable short-term benefits over the open approach, resulting in a significant decrease of 30-day mortality and faster recovery. In particular, the minimally invasive approach is associated with significantly lower incidence of overall surgical and medical postoperative complications. Based on these high-quality data, it can be stated that minimally invasive rectal surgery is safe with better short-term outcomes than open surgery. In addition, the laparoscopic approach does not jeopardize long-term survival as demonstrated by several RCTs and non-RCTs [13].

However, laparoscopic surgery for mid and lower rectal cancer is a complex and technically challenging procedure that requires a steep learning curve. It has been advocated that the use of robotic technologies might help reduce the technical difficulties encountered during the procedure and might shorten the learning curve. Standard robotic platforms, including the

Da Vinci Surgical System (Intuitive Surgical, Sunnyvale, CA, United States – www.intuitivesurgical.com), have been developed aiming at increasing the dexterity and improving the ergonomics of the surgeon. However, they are very expensive and their potential benefits in the surgical treatment of rectal cancer are not proved [14]. For instance, the RCT that compared RObotic and LAparoscopic Resection for Rectal Cancer (ROLARR) failed to demonstrate statistically significant differences in terms of rate of conversion to open surgery, intraoperative complications, early post-operative morbidity, and in the rate of positive radial margins after laparoscopic or robotic surgery [15]. Based on this evidence, the robotic technology currently available on the market that uses rigid instruments does not facilitate the procedure and does not further improve the outcomes achieved after standard laparoscopic surgery. As a consequence, the research has recently focused on the development of novel flexible devices for minimally invasive surgery [16–18]. In particular, soft and stiffness-controllable robotic technology has proved to be effectively used in different body districts, including heart, throat [19–21], brain [22, 23], and abdominal organs, through a single-port access [24, 25]. Following the experience with flexible endoscopes with enhanced features that have been used to guide tools into intra-abdominal organs through natural orifices [26, 27], highly articulated and actively guided surgical instruments have been conceived to reach the surgical site with very limited interaction with the surrounding structures [28]. The ideal tool to achieve this goal should be soft, with the ability to become stiff when considerable forces are needed for tissue or organ retraction or to accomplish surgical tasks with the end-effector [29, 30]. In 2011, these concepts and the close observation of the tentacles of the octopus led to the conception of a novel robotic platform. The project, which was funded by the European Commission within the Seventh Framework Program, started in January 2012. During the following 4 years, a robotic arm including a camera that was able to pass through a standard, commercially available 15-mm trocar was designed. This chapter aims at showing the feasibility of laparoscopic TME under the vision provided by a soft and flexible robotic camera in human cadavers.

19.2 Methods

Building up the prototype model included assembling the flexible modules and adding detecting abilities, force feedback, route control, and in addition, a user interface (UI) and complex software to empower real-time dialog of all the components. Initially, large-scale prototyped models were produced with the specific aim to verify the idea on benchtop models. A 24-mm-diameter prototype of the STIFF-FLOP arm was made, comprising various soft, pneumatically incited three-chamber sections [31]. Extra chambers were incorporated inside the sections in order to permit their hardening, utilizing an approach based on the granular jamming principle (Figure 19.1). Models with two or three sections were manufactured and some human stomach phantom models were used in order to test the framework [32]

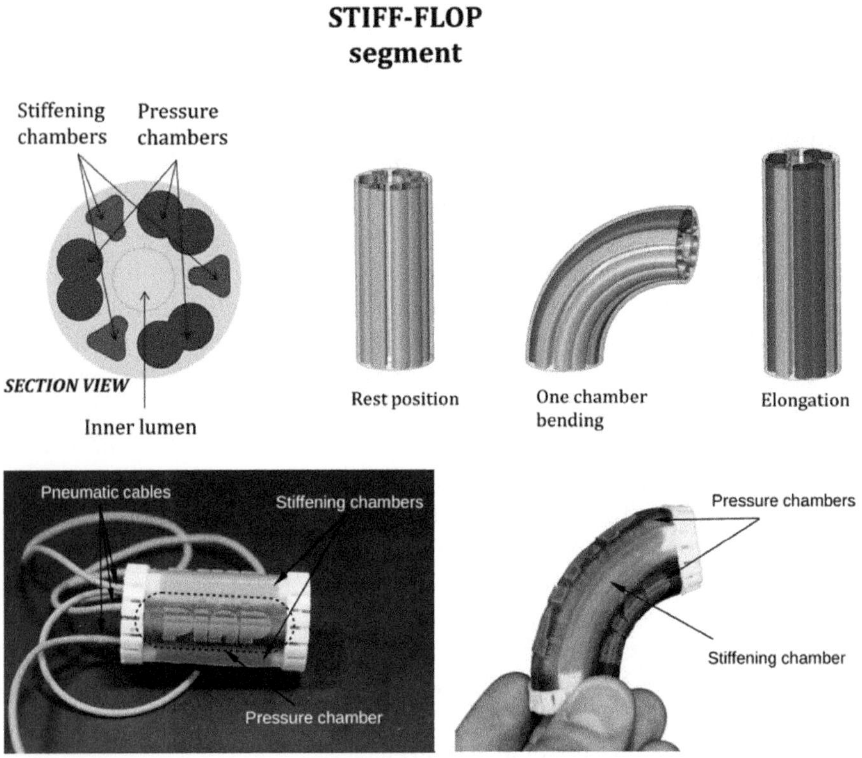

Figure 19.1 Computer model design of one STIFF-FLOP arm segment. From the left: section view of the segment showing the arrangement of the chambers (pneumatic and stiffening); segment in the rest position (no pressure is supplied to the chambers); bending of the segment due to the pressurization of one pneumatic chamber (in dark blue); elongation of the segment due to the simultaneous pressurization of all the chambers. The stiffening mechanism can be activated by controlling the level of vacuum in all the stiffening chambers (in red), once the desired position of the segment is reached.

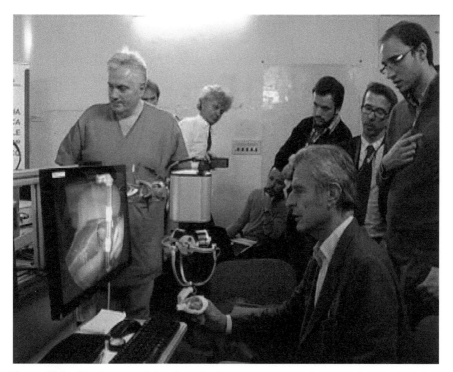

Figure 19.2 Six degrees of freedom (DoF) haptic input device, based on a Delta robot design.

(Figures 19.2 and 19.3). The STIFF-FLOP sections were activated utilizing pressure controllers, which were managed by a RoNeX-board (Shadow Robotics, London, United Kingdom). The stiffening of the three-chamber sections was controlled through valves commanded by a RoNeX-boar. When the valves were open, a vacuum is applied to the granules contained in the three chambers, which thus transform the flexible segment of the STIFF-FLOP arm into a stiff segment. Sensors were installed in the STIFF-FLOP modules to quantify interaction forces (between the robot and its environment) and the robot's setup. In this specific scenario, each section was embedded with a three-pivot force/torque (F/T) sensor and a three-degrees-of-freedom bending sensor. To increase the pose route, a laparoscopic camera, two outer sensors, and an NDI Aurora magnetic tracker (NDI International Headquarters, Waterloo, ON, Canada) were used. In order to achieve this purpose, various markers were appended at different areas along the STIFF-FLOP arm. A Schunk mechanical arm (Schunk GmbH and Co. KG,

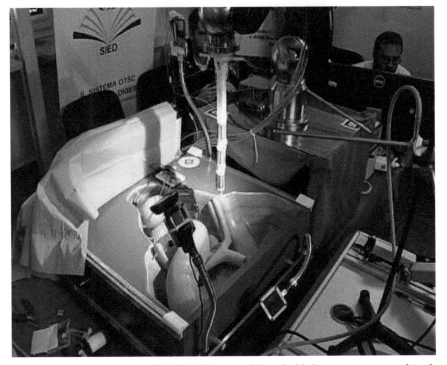

Figure 19.3 Three-segments STIFF-FLOP arm with embedded sensors, connected to the SCHUNK robot during invitro Tests done at UNITO

Hamburg, Germany) was then linked at the base of the STIFF-FLOP robotic arm, out of the body of the patient, to allow the STIFF-FLOP arm to move in and out through a trocar cannula and to position and orientate the base of the STIFF-FLOP arm as required. Contribution from the magnetic sensors guaranteed that the pivot point of the trocar and the mechanical arm were always linked. This approach guaranteed that the STIFF-FLOP arm was constantly embedded along the central longitudinal axis of the trocar cannula, allowing the pitch and yaw on the trocar access point. Control and route strategies were created to process the inverse kinematics for the extended kinematic chain of the robotic arm and the STIFF-FLOP arm continuously and in real time, thanks to the contributions from different sensors. Due to this real-time control, the operator could control the position of the tip of the STIFF-FLOP arm in the working space without the need to control the movements of each arm segment. A recently created UI, derived from a Delta robot [33], was used for moving and positioning the tip of the STIFF-FLOP arm inside the body (Figure 19.2). In the screens of the platform, the operator could see the visual feedback derived from the camera with the addition of a real-time 3D simulation demonstrating the 3D positioning of all STIFF-FLOP modules. The inputs acquired from the F/T sensors were sent back to the UI, giving to the surgeon real-time force feedback, opposing the physician's hand movements when the robot was in contact with an anatomic structure of the patient. The tip effectors of these models were outfitted with various surgical laparoscopic instruments, such as a gripper and a monopolar hook. Successful utilization of these devices in realistic situations was proven.

A thinner STIFF-FLOP model able to pass through a standard 15-mm trocar was created for *ex vivo* human setting tests. This STIFF-FLOP model has thin pneumatically activated sections, a camera that is positioned at the distal tip of the last segment, and a positioning device. The consortium effectively figured out how to downscale the entire soft robot framework to 14.3 mm diameter, equipped for being embedded into the human body through a standard trocar cannula. Downsizing the STIFF-FLOP model constrained us to renounce the detecting capacities aside from the camera; for this reason, the control of the tip of the soft robot was conceivable by independently moving the two robots' segments with the two joystick inputs. This controlling system permits each soft-robotic module to elongate along the central longitudinal axis and also to bend in any direction.

The soft robot has a 4-mm free lumen along the center of its long axis, which permits the electrical wires the passage that is required for the laparoscopic visual module situated at the tip of the robotic arm. The MD-T1003L-65 optics (Misumi, New Taipei City, Taiwan) is 12 mm long and 3.8 mm in diameter; the optics are incorporated with a light framework (four LEDs) and are connected to a computer station by a USB connector. The STIFF-FLOP camera robotic module was appended to an unbending shaft 10 mm in diameter, which was linked to the surgical table with an articulated arm with three ball-shaped joints (KLS Martin GmbH and Co. KG, Freiburg, Germany). This arm could be physically moved and adjusted in order to have the correct positioning of the robot base during the intervention. The primary goal of the test was to prove that the STIFF-FLOP robotic framework was adequate to achieve a laparoscopic TME in a human model and to assess if flexibility, softness, and dexterity of the STIFF-FLOP visual module may represent an advantage when compared to standard unbending laparoscopic tools.

19.3 Operative Technique

The feasibility and effectiveness of the 14-mm robotic STIFF-FLOP camera during a laparoscopic TME on a human cadaver model were tested at the Institute for Medical Science and Technology (IMSaT), Dundee, Scotland. The robotic system, including the software and the robotic STIFF-FLOP camera, was installed by the engineers. The robot-assisted laparoscopic TMEs were performed on two cadavers that were embalmed following the Thiel method. According to this technique, salt compounds mixed to very low quantities of volatile formaldehyde and formalin are used to fix tissues, guaranteeing excellent antimicrobial properties, keeping a life-like flexibility of the different segments of the cadaver, and preserving the natural color of muscles, viscera, and vessels, with no detectable odor.

The suitability of the two human cadaver models had been tested by surgeons in a previous lab session. After positioning and safely securing each cadaver to a dedicated operating table, the entire instrumentation was checked. Then, a 10-mm trocar was placed trans-umbilically to get access to a 30° camera and to create a stable pneumoperitoneum. The intra-abdominal organs and the abdominal wall compliance to the pneumoperitoneum were assessed. The following step included the placement of three 5-mm trocars in the right flank, right iliac fossa, and left flank. Then, the surgeon carefully checked the thickness of both bowel and mesocolon. Lastly, the surgeon mobilized the left colon from the abdominal wall and checked the feasibility of mesenteric vessel dissection.

The surgical team included three surgeons. Before starting the laparoscopic TME, the cadaver was placed on and secured to the operating table, and all operative tools were checked. Four trocars were used: one 15-mm trocar was positioned on the midline about 2 cm above the umbilicus, and the other three 5/12-mm trocars were positioned in the right flank, left flank, and right iliac fossa, respectively. The consistence of bowel and mesocolic fatty tissue was then checked under the vision of a standard 10-mm 30° laparoscopic camera positioned through the median trocar. Then, a fifth 10-mm trocar was inserted in the left upper quadrant, posterior to the median trocar, aiming at achieving an overview by a standard 10-mm 30° laparoscopic camera and introducing under direct vision the flexible STIFF-FLOP camera (Figure 19.4). The entire laparoscopic TME was followed on two screens: one monitor was connected to the standard laparoscopic camera and the second to the STIFF-FLOP optic. The surgical procedure was entirely

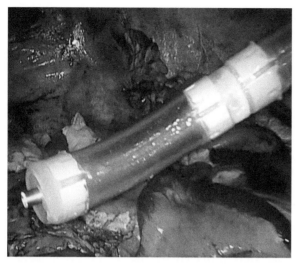

Figure 19.4 Flexible STIFF FLOP camera during cadaver tests.

recorded at a standard 24-frame per second rate for subsequent review and critical analysis.

During the first step of the surgical procedure, the medial dissection of the sigmoid mesocolon was performed using standard laparoscopic tools. Then, the inferior mesenteric vessels were identified, dissected, clipped, and divided. The STIPP-FLOP camera helps the surgeon to clearly recognize the inferior mesenteric vessels and the autonomic nerves that were successfully spared. The operation continued with the identification of the iliac vessels and both ureters, and with the posterior dissection of the mesorectum in the presacral avascular plane down to the pelvic floor. Afterwards, the surgeon completed the mesorectal excision laterally on both right and left rectal sides. Lastly, the anterior dissection of the mesorectum was completed after identifying the Denonvillers' fascia and preserving the seminal vesicles and the lateral pelvic nerves. At the end of the procedure, the rectum was circumferentially mobilized and prepared for transection. The quality of the dissection was evaluated by assessing the integrity of the mesorectal fascia.

19.4 Results

A team including three surgeons performed both laparoscopic TME procedures. One of the three surgeons used a dedicated joystick to orient the STIFF-FLOP camera by means of an X-BOX controller for each robotic module, thus allowing bending movements in the two axes and translation. The elongation and retraction of the system were both assisted by an engineer. One surgeon performed the operation, while the third surgeon drove the standard 30o laparoscopic camera. The STIFF-FLOP camera was controlled only by the visual feedback provided by the camera, without the help of any other sensor.

As the first step of the laparoscopic TME, the sigmoid mesocolon was medially dissected, and both inferior mesenteric artery and vein identified and divided under the direct view given by the STIFF-FLOP camera. The STIFF-FLOP camera also clearly showed the autonomic nerves, avoiding the risk of nerve injuries. After the completion of the sigmoid mesocolon dissection, the surgeon moved toward the pelvis to perform the mesorectal excision under the direct vision provided by the STIFF-FLOP camera. The mesorectal dissection was first started in the posterior avascular plane: the enhanced and stable view provided by the STIFF-FLOP camera and its flexibility facilitated this surgical step by allowing the surgeon to follow the sacral curve very closely and perform the dissection precisely. After reaching the pelvic floor posteriorly, the surgeon completed the mesorectal excision laterally and anteriorly. Also, this step of the laparoscopic TME was not difficult to carry out thanks to the magnification of the view of the surgical field that was achieved by using the flexible STIFF-FLOP camera. At the end of the operation, the mesorectal excision was complete and the mesorectal fascia was intact. The overall operative time was 165 min for the first case and 145 min for the second case. There were no intra-operative adverse events. From a technical point of view, no problems were encountered. At the beginning of each procedure, the STIFF-FLOP robotic arm was inserted in the abdomen through a 15-mm trocar without any difficulties; the surgeons did not report any movement limitation during the navigation forward, backward, and laterally. The cleaning of the STIFF-FLOP camera was required only twice during each operation and was performed by taking out the arm from the abdomen as is usually done with the standard rigid laparoscopic camera. Only a few minutes of training were necessary to achieve the adequate ability to manipulate the dual-joystick input device. Intra-abdominal navigation was facilitated by a double check looking at the screen connected to the standard laparoscopic camera.

The three surgeons performed the same surgical tasks on two human cadavers showing the friendly use and robustness of this robotic camera for several hours.

19.5 Discussion

The minimally invasive approach to extra-peritoneal rectal cancer is safe and feasible, even though it is technically demanding. Despite the obvious advantages from the patient's perspective [10, 11], the laparoscopic approach has not gained wide acceptance and to date it is not routinely adopted. The use of robots was conceived to improve the surgical outcomes in rectal cancer patients. However, the currently available literature does not show clear benefits from the robotic technology over the standard laparoscopy, in terms of both ergonomics of the surgeon and outcomes of the patient [13]. Major efforts have been made to improve surgical training for both laparoscopic and robot-assisted surgery. We feel that the implementation of the new idea of flexible robotic arms on top of the existing robotic technology may be key to reducing the "human" factor that has a major impact on the outcomes of surgical procedures performed using current "rigid" technology.

For these reasons, in 2011, we conceived a new soft and flexible robot that was characterized by the ability to easily reach narrow spaces, including the pelvis. During this 4-year project, we were able to develop a modular technology, made of soft and flexible modules, able to bend under pneumatic swelling of dedicated chambers, and also to become stiff when required using the granular jamming technology of committed chambers incorporated in the STIFF-FLOP arm prototypes. The UI that controls all the movements consist of two separate joysticks derived from a modified Delta robot. The STIFF-FLOP arm is designed in order to permit unconstrained and free independent movements of every module in any direction, bending from the major axis of the module, accomplishing an extensive variety of movements and a large workspace. Moreover, lengthening can also be accomplished. The UI controller leaves the operator the control in real time of the spatial position of the end-effector tip. It assures the positioning of the system by setting the correct spatial coordinates of where the tip of the arm should move to and its orientation. In the meantime, in order to get the desired spatial position and orientation of the arm, the arm will move, like an octopus tentacle, driven from a software algorithm that will compare in real time the required position, given from the UI, and the real position and automatically route the arm in the best way to reach the target position and orientation. This routing is made possible by the several sensors set along the arm, which permit a fine control of the position of every module, and additionally, thanks also to the force sensors that are able, in conjunction with the visual output, to give a haptic feedback to the operator through the UI input device. These unique characteristics make this new robotic platform extremely innovative compared to the currently available flexible surgical tools. Indeed, the flexibility of both robotic and laparoscopic tools is restricted to the tip of the device, which keeps a rigid stem that does not allow following the curved surfaces inside the human body. The presence of a modular, stiff, but, at the same time, soft and flexible arm when necessary, might allow easy reach of any target inside the abdomen or the chest.

Theoretically, these characteristics should exceed the majority of the drawbacks of the currently available surgical robots, which are limited by a rigid architecture. With a specific end-goal of being able to insert the arm into a standard trocar for the human cadaver study, we decided to cut off the detecting abilities in order to be able to scale-down the arm and be able to test the soft and flexible arm concept in a first sensorless fashion. For testing, the miniaturized STIFF-FLOP robot was connected to a rigid shaft in turn attached to the operative table by means of an anthropomorphic arm. A double joystick was used to control the entire system. We tested this robotic tool on two human cadavers, aiming at assessing and demonstrating the feasibility, effectiveness, and adequacy of its geometry and function.

We were able to complete a TME procedure with the help of the STIFF-FLOP camera in two consecutive human cadavers by using standard laparoscopic rigid instruments. The operative time for both TMEs was shorter than 180 min and neither complications nor technical issues were experienced. We appreciated the ability of the STIFF-FLOP arm to enter the pelvis getting very close to the sacral bone and the lateral walls of the rectum, thus allowing a very precise mesorectal excision that is otherwise very difficult during a laparoscopic TME. The ability to provide a magnified and close view of this low and very limited surgical field has demonstrated the correct concept and geometry of the robotic tool.

We have shown that a TME procedure performed under the direct vision of a soft and flexible camera is feasible and safe in human cadaver models. However, further cadaver test and possibly clinical studies are needed to confirm these preliminary findings and studies comparing the STIFF-FLOP camera with the last version of 3D high-definition cameras with flexible tip are awaited to better define the possible impact of this robotic technology in the clinical practice.

The STIFF-FLOP technology has some strengths and weaknesses. Strengths include the magnified vision of the surgical field with good image quality and the steady control of the camera movements by using the dedicated joystick. Weaknesses include the reduced intuitiveness of the control of the device, which might be improved by adopting the inverse kinematics-based approach that was used in the large STIFF-FLOP arm prototype. The insertion of sensors in the prototype used during the TME procedures and the use of computing inverse kinematics in real time will permit the employment of the modified Delta robot to control both orientation and position of the robotic system. Consequently, the navigation of the robotic camera will improve and become more user friendly. During the project period, we have also developed a new robotic prototype with an embedded gripper on the tip and a monopolar coagulator that will let us shortly attempt a complete laparoscopic procedure by means of soft and flexible robotic tools.

19.6 Conclusions

This chapter reports the first two cases of laparoscopic surgical procedures performed with the assistance of a flexible and soft robot. Based on this preliminary experience, we feel that the optimized vision of the surgical field along with the flexibility of the robotic camera might significantly help the surgeon to perform a technically demanding surgical procedure, such as laparoscopic TME, in a very precise way. However, further cadaver tests are required to prove the safety of this very promising technology before suggesting clinical trials.

References

[1] Lacy, A. M., Garcia-Valdecasas, J. C., Delgado, S., Castells, A., Taura, P., Pique, J. M., et al. (2002). Laparoscopy-assisted colectomy versus open colectomy for treatment of non-metastatic colon cancer: a randomised trial. *Lancet* 359, 2224–2229.
[2] Guillou, P. J., Quirke, P., Thorpe, H., Walker, J., Jayne, D. G., Smith, A. M., et al. (2005). Short-term endpoints of conventional versus laparoscopic-assisted surgery in patients with colorectal cancer (MRC CLASICC trial): multicentre, randomised controlled trial. *Lancet* 365, 1718–1726.
[3] Laparoscopically assisted colectomy is as safe and effective as open colectomy in people with colon cancer. Abstracted from: Nelson, H., Sargent, D., Wieand, H. S., et al. (2004). For the Clinical Outcomes

of Surgical Therapy Study Group. A comparison of laparoscopically assisted and open colectomy for colon cancer. N Engl J Med 2004; 350: 2050-2059. *Cancer Treat Rev*. 30, 707–709.

[4] Veldkamp, R., Kuhry, E., Hop, W. C., Jeekel, J., Kazemier, G., Bonjer, H. J., et al. (2005). Laparoscopic surgery versus open surgery for colon cancer: short-term outcomes of a randomised trial. *Lancet Oncol*. 6, 477–484.

[5] Heald, R. J., Moran, B. J., Ryall, R. D., Sexton, R., and MacFarlane, J. K. (1998). Rectal cancer: the Basingstoke experience of total mesorectal excision, 1978–1997. *Arch. Surg*. 133, 894–899.

[6] Ries, L. A., Wingo, P. A., Miller, D. S., Howe, H. L., Weir, H. K., Rosenberg, H. M., et al. (2000). The annual report to the nation on the status of cancer, 1973–1997, with a special section on colorectal cancer. *Cancer* 88, 2398–2424.

[7] van der Pas, M. H., Haglind, E., Cuesta, M. A., Furst, A., Lacy, A. M., Hop, W. C., et al. (2013). Laparoscopic versus open surgery for rectal cancer (COLOR II): short-term outcomes of a randomised, phase 3 trial. *Lancet Oncol*. 14, 210–218.

[8] Bonjer, H. J., Deijen, C. L., Abis, G. A., Cuesta, M. A., van der Pas, M. H., de Lange-de Klerk, E. S., et al. (2015). A randomized trial of laparoscopic versus open surgery for rectal cancer. *N. Engl. J. Med*. 372, 1324–1332.

[9] Fleshman, J., Branda, M., Sargent, D. J., Boller, A. M., George, V., Abbas, M., et al. (2015). Effect of Laparoscopic-Assisted Resection vs Open Resection of Stage II or III Rectal Cancer on Pathologic Outcomes: The ACOSOG Z6051 Randomized Clinical Trial. *JAMA* 314, 1346–1355.

[10] Stevenson, A. R., Solomon, M. J., Lumley, J. W., Hewett, P., Clouston, A. D., Gebski, V. J., et al. (2015). Effect of Laparoscopic-Assisted Resection vs Open Resection on Pathological Outcomes in Rectal Cancer: The ALaCaRT Randomized Clinical Trial. *JAMA* 314, 1356–1363.

[11] Arezzo, A., Passera, R., Scozzari, G., Verra, M., and Morino, M. (2013). Laparoscopy for rectal cancer reduces short-term mortality and morbidity: results of a systematic review and meta-analysis. *Surg. Endosc*. 27, 1485–1502.

[12] Arezzo, A., Passera, R., Scozzari, G., Verra. M., and Morino, M. (2013). Laparoscopy for extraperitoneal rectal cancer reduces short-term morbidity: Results of a systematic review and meta-analysis. *United European Gastroenterol. J*. 1, 32–47.

[13] Arezzo, A., Passera, R., Salvai, A., Arolfo, S., Allaix, M. E., Schwarzer, G., et al. (2015). Laparoscopy for rectal cancer is oncologically adequate: a systematic review and meta-analysis of the literature. *Surg. Endosc.* 29, 334–348.

[14] Vallribera Valls, F., Espin Bassany, E., Jimenez-Gomez, L. M., Ribera Chavarria, J., and Armengol Carrasco, M. (2014). Robotic transanal endoscopic microsurgery in benign rectal tumour. *J Robot. Surg.* 8, 277–280.

[15] Jayne, D., Pigazzi, A., Marshall, H., Croft, J., Corrigan, N., Copeland, J., et al. (2017). Effect of robotic-assisted vs conventional laparoscopic surgery on risk of conversion to open laparotomy among patients undergoing resection for rectal cancer: the rolarr randomized clinical trial. *JAMA* 318, 1569–1580.

[16] Jiang, A., Ataollahi, A., Althoefer, K., Dasgupta, P., and Nanayakkara, T. (2012). "A Variable Stiffness Joint by Granular Jamming," in *Proceedings of the Asme International Design Engineering Technical Conferences and Computers and Information in Engineering Conference 2012*, Vol. 4, Pts a and B. 267.

[17] Jiang, A., Xynogalas, G., Dasgupta, P., Althoefer, K., and Nanayakkara, T. (2012). "Design of a variable stiffness flexible manipulator with composite granular jamming and membrane coupling," in *Proceedings of the IEEE/RSJ International Conference on Intelligent Robots and Systems (IROS)*, 2922–2927.

[18] Stilli, A., Wurdemann, H. A., and Althoefer, K. (2014). "Shrinkable, stiffness-controllable soft manipulator based on a bio-inspired antagonistic actuation principle," in *Proceedings of the 2014 Ieee/Rsj International Conference on Intelligent Robots and Systems (Iros 2014)*, 2081.

[19] Degani, A., Choset, H., Wolf, A., and Zenati, M. A. (2006). "Highly articulated robotic probe for minimally invasive surgery," in *Proceedings of the 2006 IEEE International Conference on Robotics and Automation, 2006 (ICRA 2006)*, 4167.

[20] Wei, W., Xu, K., and Simaan, N. (2006). "A compact two-armed slave manipulator for minimally invasive surgery of the throat," in *Proceedings of the The First IEEE/RAS-EMBS International Conference on Biomedical Robotics and Biomechatronics, 2006 (BioRob 2006)*, 1190.

[21] Bajo, A., Dharamsi, L. M., Netterville, J. L., Garrett, C. G., and Simaan, N. (2013). "Robotic-Assisted Micro-Surgery of the Throat: the Trans-Nasal Approach," in *Proceedings of the IEEE International Conference on Robotics and Automation (ICRA)*, 232–238.

[22] Burgner, J., Swaney, P. J., Lathrop, R. A., Weaver, K. D., and Webster, R. J. (2013). Debulking from within: a robotic steerable cannula for intracerebral hemorrhage evacuation. *IEEE T Bio-Med. Eng.* 60, 2567–2575.

[23] Ho, M. Y., McMillan, A. B., Simard, J. M., Gullapalli, R., and Desai, J. P. (2012). Toward a Meso-Scale SMA-Actuated MRI-Compatible Neurosurgical Robot. *IEEE T. Robot.* 28, 213–222.

[24] Bajo, A., Goldman, R. E., Wang, L., Fowler, D., and Simaan, N. (2012). "Integration and Preliminary Evaluation of an Insertable Robotic Effectors Platform for Single Port Access Surgery," in *Proceedings of the IEEE International Conference on Robotics and Automation (ICRA)*, 3381–3387.

[25] Shang, J., Noonan, D. P., Payne, C., Clark, J., Sodergren, M. H., and Darzi, A., et al. (2011). "An articulated universal joint based flexible access robot for minimally invasive surgery," in *Proceedings of the IEEE International Conference on Robotics and Automation (ICRA)*, 1147–1152.

[26] Bardaro, S. J., and Swanstrom, L. (2006). Development of advanced endoscopes for Natural Orifice Transluminal Endoscopic Surgery (NOTES). *Minim Invasive Ther. Allied Technol.* 15, 378–383.

[27] Phee, S. J., Ho, K. Y., Lomanto, D., Low, S. C., Huynh, V. A., Kencana, A. P., et al. (2010). Natural orifice transgastric endoscopic wedge hepatic resection in an experimental model using an intuitively controlled master and slave transluminal endoscopic robot (MASTER). *Surg. Endosc.* 24, 2293–2298.

[28] Loeve, A., Breedveld, P., and Dankelman, J. (2010). Scopes too flexible... and too stiff. *IEEE Pulse.* 1, 26–41.

[29] Mahvash, M, and Dupont, P. E. (2011). Stiffness control of surgical continuum manipulators. *IEEE T. Robot.* 27, 334–345.

[30] Goldman, R. E., Bajo, A., and Simaan, N. (2014). Compliant motion control for multisegment continuum robots with actuation force sensing. *IEEE T. Robot.* 30, 890–902.

[31] Fras, J., Czarnowski, J., Macias, M., Glowka, J., Cianchetti, M., and Menciassi, A. (2015). "New STIFF-FLOP module construction idea for

improved actuation and sensing," in *Proceedings of the IEEE International Conference on Robotics and Automation (ICRA)*, 2901–2906.

[32] Małota, Z., Nawrat, Z., and Sadowski, W. (2014). Symulatory chirurgii małoinwazyjnej. *Eng. Biomat.* 17, 2–11.

[33] Miller, K., and Clavel, R. (1992). The lagrange-based model of delta-4 robot dynamics. *Robotersysteme* 8, 49–54.

Index

About the Editors

Jelizaveta Konstantinova is a Roboticist at The Centre for Advanced Robotics @ Queen Mary University of London. She has received a professional bachelor degree and engineer qualification in Mechatronics from Riga Technical University, Riga, Latvia, in 2010. Further on, she joined King's College London in 2010 and obtained MSc degree in Robotics with the focus on Medical Robotics. From 2011 to 2015, Dr Konstantinova was working towards her PhD degree in Robotics, and from 2015 she was a research associate on EU FP7 project SQUIRREL within the Centre for Robotics Research, King's College London. In 2016 she has moved to Queen Mary University of London as a Postdoctoral Research Assistant. Her research focus is on haptic perception, development of tactile technologies, medical robotics, sensory data fusion and learning for robotics, as well as on design and modelling of soft and flexible robotic manipulators. She is interested to learn the perception capabilities from living organisms and to apply it to robotics, as well as to study the ways we should interpret tactile information obtained with robotic sensory system.

Helge Wurdemann is a Roboticist and Lecturer at University College London with a research focus on the design and application of bio-inspired, soft and stiffness-controllable robotic structures and haptic interfaces. Dr Wurdemann graduated in Electrical Engineering (Mechatronics) from the Leibniz University of Hanover, Germany in 2008 and completed a PhD in Mechatronics and Robotics at King's College London, UK in 2008. He started to work as a post-doctoral research associate and project manager for the EU project STIFF-FLOP coordinating research activities of the multi-disciplinary consortium of 12 partners with a total budget of € 7.35m. At King's College London, Dr Wurdemann was Head of the Soft Robotics Lab and Haptics Lab within the Centre for Robotics Research. In January 2016, he then joined the Department of Mechanical Engineering at University College London and established the Soft Haptics and Robotics Lab within the Biomechanical Research Group. Dr Wurdemann leads the robotics activities for UCL Mechanical Engineering and represents the Department within the UCL Robotics Institute. Further, Dr Wurdemann is a member of the UCL Institute of Healthcare Engineering. His research has resulted in over 50 publications in high-impact journals and presentations at international premium conferences in robotics.

Ali Shafti has a Ph.D. in Robotics from King's College London, UK. He obtained his B.Sc. and M.Sc. in Microelectronics Engineering at Shahid Beheshti University and Amirkabir University of Technology, Tehran, Iran respectively. As part of his PhD, Ali was involved in the STIFF-FLOP project among others, where he explored human-robot interaction/collaboration, wearable technologies, and biomedical instrumentation for human factors. He is currently a Postdoctoral Research Associate at the Brain and Behaviour Laboratory and the Data Science Institute, Imperial College London, where he performs research on human-robot collaborative control, through human behaviour analytics and human/robot in-the-loop machine learning techniques.

Ali Shiva received his BSc from Tehran University and his MSc from the University of Arizona in Mechanical Engineering. Ali has worked for about ten years as a design engineer and engineering coordinator in the field of gas, oil and petrochemical plants. Since October 2014, Ali has been working as a PhD student at Centre of Robotics Research (CoRe) at King's College London. His main areas of interest are Robotics, Dynamics, Multi Body Dynamics, and Control. Ali is currently working on modelling, simulation and control of pneumatically actuated soft continuum manipulators. In addition, he has been part of the Kings College team on the EU projects "STIFF-FLOP" and "FourbyThree", and is currently a Research Associate at Queen Mary University of London. He is also a visiting researcher at Imperial College London.

Kaspar Althoefer is an electronics engineer, leading research on Robotics at Queen Mary University of London. After graduating with a degree in Electronic Engineering from the University of Technology Aachen, Germany, and obtaining a PhD in Robot Motion Planning from Kings College London, he joined the Kings Robotics Group in 1996 as a Lecturer. Made a Senior Lecturer in 2006, he was promoted to Reader and Professor in 2009 and 2011, respectively. In April 2016, he joined Queen Mary as full Professor of Robotics Engineering. His current research interests are in the areas of robot autonomy, soft robotics, modelling of tool-environment interaction dynamics, sensing and neuro-fuzzy-based sensor signal classification with applications in robot-assisted minimally invasive surgery, rehabilitation, assistive technologies and human-robot interactions in the manufacturing environment. Prof Althoefer has authored/co-authored more than 200 peer-reviewed papers. The majority of his journal papers (over 60%) are in the top journals of the field, including top transactions and journals of the IEEE and ASME and proceedings of the leading national learned societies in the field, IMechE and IET. He is named inventor on five patent applications.